ENGINEERING SUSTAI
ON EARTH

Climate scientists have clarified the main causes of climate change, and the tight timescale within which humans must change their behaviour and implement effective solutions, wherever they are needed across the world. This book uncovers many of the powerful actions and uses them effectively to achieve sustainable human life, of improved quality, in a way that is affordable out of earned income for all humans, wherever they live.

The ultimate solution to climate change lies not just in doing and consuming less but does instead entirely revolve around our ability to "out-innovate" the problem. John F. Coplin, CBE, FREng, FCGI, has had a long and distinguished career in engineering and has operated and advised at all levels from heads of state, company chairs, engineering directors, government advisory boards, and on the shop floor. He is perfectly placed to take a wide-ranging approach, applying modern design and innovative engineering at a systemic level in order to provide novel approaches that will have a far-reaching impact on reversing humankind's impact on the planet. His projections and solutions are based on facts, reasonable calculations, and science learnt from nature. Unafraid to challenge current thinking, John looks at solutions across multiple sectors, including aviation, cars and domestic local transport, clean and renewable energy, food and agriculture, and housing and communities, and describes the particular potential of hydrogen as a fuel.

The book is written in a language for all. It is small enough to be used as a practical guide to where some of the most useful improvements are to be found and as a way to start important conversations.

John F. Coplin, aeronautical engineer, chief designer of Rolls-Royce's RB211 aeroengine. During the 1990s, he was UK science and technology adviser to the Indonesian President. Previously a visiting professor on engineering design at Oxford University and Imperial College and Associate Fellow in Design Engineering at the University of Warwick.

ENGINEERING SUSTAINABLE LIFE ON EARTH

Alleviating Adverse Climate Change Through Better Design

John F. Coplin

Edited by Dave Coplin

Routledge
Taylor & Francis Group

LONDON AND NEW YORK

First published 2022
by Routledge
2 Park Square, Milton Park, Abingdon, Oxon OX14 4RN

and by Routledge
605 Third Avenue, New York, NY 10158

Routledge is an imprint of the Taylor & Francis Group, an informa business

© 2022 John F. Coplin

British Library Cataloguing-in-Publication Data
A catalogue record for this book is available from the British Library

Library of Congress Cataloging-in-Publication Data
Names: Coplin, John F., 1934- author.
Title: Engineering sustainable life on earth: alleviating adverse climate change through better design / John F. Coplin.
Description: Abingdon, Oxon; New York, NY: Routledge, 2022. | Includes bibliographical references and index. |
Identifiers: LCCN 2021012794 (print) | LCCN 2021012795 (ebook) | ISBN 9781032044958 (hbk) | ISBN 9781032044965 (pbk) | ISBN 9781003193470 (ebk)
Subjects: LCSH: Sustainable development. | Climatic changes. | Renewable energy sources.
Classification: LCC HC79.E5 C658 2022 (print) | LCC HC79.E5 (ebook) | DDC 363.738/74–dc23
LC record available at https://lccn.loc.gov/2021012794
LC ebook record available at https://lccn.loc.gov/2021012795

ISBN: 978-1-032-04495-8 (hbk)
ISBN: 978-1-032-04496-5 (pbk)
ISBN: 978-1-003-19347-0 (ebk)

DOI: 10.4324/9781003193470

Typeset in Joanna
by Deanta Global Publishing Services, Chennai, India

This book is dedicated to all my grandchildren, Max, Harry, John, Ellie, and James Coplin.

May your lives be sustainable and happy.

To Dr. D. Smith
Headmaster

May your future be bright
and sustainable

John Coplin
Bablake 1945 – 1953.

Nov. 2021

CONTENTS

FOREWORD

This book shows how lessons learnt from leadership in design engineering can help avoid a climate disaster causing an extinction event. The extent of the problem and the magnitude of the action required to avert the disaster require us to do more than simply "consume less" but instead highlight the desperate need for us to "out-innovate" the problem through the application of engineering principles, innovation, and research. This will create opportunities that will help recover the environment without having to compromise on too many of the aspects of our lives that currently contribute towards the death of our planet.

Most people are no longer in any doubt that humans are responsible for climate change. This problem grows exponentially as more and more of the Earth's population seek to achieve an aspirational lifestyle that paradoxically provides valuable progress in terms of health and societal outcomes, but at the same time, causes increasing damage to the environment.

Through this book, I intend to use my experience in aerospace design engineering to show how new developments and innovations, along with appropriate behavioural changes, can not only avert an impending climate disaster but create a sustainable approach to life that enables us to continue to travel, develop, and grow but without doing so at the cost of our planet.

Wherever possible, I have used publicly available data at the time of publishing to help demonstrate what needs attention and to illustrate what can be done, but I recognize that these numbers are changing quickly as we learn more and our behaviour changes. I'm hopeful you'll find my logic useful, even as the numbers evolve.

We will look at many of the steps that must be taken to avoid a disaster, noting that there are strong interactions linking them all as a reset of our human activity achieves better health and wealth outcomes for all, with increased employment in a larger global market.

However, in order to be effective, we all need to be involved and committed to making the changes required. Further still, the changes must deliver tangible benefits quickly to attract the necessary investment.

Thankfully, everything we need to make this result possible already exists, all we really need now is the action to make it happen as it is only through our collective actions that we can achieve sustainable life in a way that avoids a climate disaster.

As a design engineer, I have spent much of my career with some outstanding colleagues and friends, working within difficult constraints to develop safe solutions for complex systems that can help to move society forwards. I know first-hand what is possible when the right people come together united by a common goal.

Through these experiences, I think three fundamental lessons will help guide design engineers through the challenge that lies ahead in order to engineer sustainable life on Earth. They are simply to understand that:

1. Innovation is essential to sustain the growing population on Earth but innovation carries risk.
2. Those who get things wrong while innovating are usually best placed to get those things right.
3. There is more joy in recovering from a problem than is lost by causing it.

Even in the face of such an intimidating and complex problem, I remain optimistic and enthused by the prospect of what I know engineers can achieve. With the right support and resources and with the will of all the world's citizens behind them, I know they will not just help us avoid disaster, they will also help ensure that all life can flourish in a way that does not deplete our precious environment, but instead, restores and renews it.

John F. Coplin
London, January 2021

Part 1

THE BIG RESET

DOI: 10.4324/9781003193470-1

1

ENGINEERING EXTINCTION?

DOI: 10.4324/9781003193470-2

As chief designer of engines that power many of the world's wide-bodied jet aircraft, I have a view on achieving safe solutions. Aeroengines are complicated machines demanding peak performance, close to fine limits, from many thousands of parts. They propel the aircraft at just the right speed for the load and atmosphere they fly through. They provide the power to all the aircraft controls. They deliver clean cabin air at just the right pressure and temperature with the right amount of oxygen. They provide heat and refrigeration to feed and hydrate everyone on board. They also deliver much of the braking force required on landing to bring the aircraft to rest. The designer must make sure that when anything unexpected occurs, there is a safe solution that causes neither harm nor alarm. When the unexpected happens, the designer must provide evidence to airline bosses that everything is safe, such that the airline can continue to generate its revenue. This is but a small-scale model of why it takes a systemic approach to achieve sustainable life for all humans on Earth. The surgeons that saved my life, after five component failures in my heart, used the same rules for safety in achieving a good outcome. Years of experience as a visiting professor in design engineering at the University of Oxford and Imperial College, London, plus eight years in the Far East as an adviser to the Government of the Republic of Indonesia, reporting to the Third President Professor Dr Ing BJ Habibie, has given me deep insight to the industries of the ASEAN nations. I have worked with leading-edge high technology companies in many of the most industrialized countries of the world. All this provides a breadth of experience from which to illuminate many of the paths needed to avert a climate disaster by securing a better life for all humans in a sustainable form.

This book is a story about engineering leadership helping to prevent a climate disaster, while increasing employment globally, improving life quality for millions, and generating revenue for nations to take better care of the poorest in every country. Since many of the new jobs will arise from actions to avoid a climate disaster, engineering a sustainable future means we can both save the planet and the economy.

The UK initiated, and benefitted greatly from, the Industrial Revolution, but sadly at that time was unaware of the climate problem that was being created as a by-product of such progress. Since then, industrialization on a global scale has accelerated the likelihood of an imminent climate disaster, thereby focusing the need for urgent global action to prevent this from happening. Climate scientists have explained the importance of greenhouse gases resulting from human activity and they have also shown the dangerous levels of these gases persisting in Earth's climate, highlighting the urgency of powerful actions to reduce further additions to greenhouse gases to net zero and removing as much of them as possible from the atmosphere. Many actions are needed, and most interact with each other.

We cannot address individual actions, instead, a total system approach is essential.

As a leading engineering designer, I spent over three decades helping to make quieter and more efficient commercial aircraft. Our success has been swamped by the masses of people all over the world now able to afford and enjoy aspects of a more affluent life. Air travel is a great example of this, between 1970 and just before the start of the pandemic, air travel passengers doubled from c. 0.3 to 4.5 billion, representing an increase of 1400% in just 50 years.[1] Advances in all industries have raised living standards and greatly increased spending power for billions of humans. Much of this is spent on products and activities such as the massive increase in air travel that make the prospect of climate disaster loom ever closer.

Climate scientists have identified the main causes of climate change. In the Royal Institution Christmas Lectures for 2020, Professor Chris Jackson, who holds the Chair in Sustainable Geoscience at Manchester University, looked back over deep geological time, charting Earth's climate from hothouse to icehouse and back again and revealing how each of the extremes had led to extinction events.[2] What is so disturbing is the current trend in global warming is rising much faster than any of the changes millions of years back in geological time.

Furthermore, this is happening now when the Earth can no longer accommodate the accumulation of damage caused over the last two centuries since the start of the Industrial Revolution. This emphasizes the extreme urgency for strong corrective measures to drastically slow the actions causing climate damage, and the need to remove greenhouse gases persisting in Earth's atmosphere.

Climate scientists have made it clear that greenhouse gases generated by human activity are to blame and of these, carbon dioxide (CO_2) is the dominant gas causing concern, augmented by methane, oxides of nitrogen, and fine particulates.

Human activity has adversely affected the role of the interconnected oceans in controlling the balance of greenhouse gases in the atmosphere. Earth's atmosphere is also dynamic, allowing greenhouse gases to have an impact on all of it. In his book, "Sustainable Energy",[3] the late David JC MacKay explained where the damage comes from and who needs to change, noting that it is the behaviour of citizens in the most industrialized nations that must make the biggest changes.

Humans have, and continue to emit far too much greenhouse gas. Moreover, humans continue to destroy the very resources, such as trees, (rain)forests, and peat bogs, that form the Earth's natural way of removing these harmful gases. In a sense, engineers led the world to this position, with no one thinking through what damage would arise as more and more of our global population became active consumers.

Thankfully, today's leading engineers recognize that we must enable the populations of all nations to remove all climate-damaging activities by illuminating safe ways of achieving that. Success requires the consent of everyone, using changes that make business as well as environmental sense. We can see many powerful solutions already, but we need to stimulate further innovations as populations, and human spending power, continue to rise.

Climate change is now at the tipping point where human life gets increasingly difficult and is heading towards disaster. Climate scientists have identified that historic and current high releases of greenhouse gases and fine particulates must be massively reduced. But we cannot address these two issues alone, we must also restore, and supplement, many of nature's natural processes for restoring the right levels of greenhouse gases required for sustainable human life, while still improving the quality of life for all.

Human behaviour, with much of it instigated by the UK's catalytic role in sparking the global Industrial Revolution, is responsible for the dire nature of our situation. Collectively, humans know much of what is needed. It remains to get the many necessary solutions into global mass markets in a way that benefits all citizens.

There remains great scope for further innovation, and refinement, and learning from new experiences as the world population transitions to a new and better way of living and working. I want my readers to be encouraged by the big improvements we know we can achieve, and I want to stimulate people to find, then implement, further developments that make life better, as well as making human life on Earth sustainable.

First, we must be clear about what problems we are required to solve. We must arrest the runaway climate, well before climate change is a disaster from which we cannot recover. This requires us to create better ways of living and working while satisfying the demand for well-paid jobs that generate the revenue to meet the needs of the new businesses, and to support all citizens with a good quality of life. This can be achieved using innovation and new technology productively in a total system approach to enable communities to win an increased share of global investment and to prosper from new products and services within the global market. Every change requires many complementary changes.

Developments must deliver tangible benefits not just over the long term, but increasingly in the short term if they are to attract the right level of attention and investment from both the public and private sectors. We know that the transition from proof of potential to the mass market is expensive, so there is a real need for governments to share in the risk, because without that, innovations will not reach the level required to generate tax revenues sufficient to support a satisfactory lifestyle for all, and borrowing will become too expensive.

Increased self-reliance is required, both at a national and personal level. This can still be compatible with high levels of global trading needed to ensure mutual understanding between nations. Avoiding a climate disaster requires increased national self-reliance. Self-reliance can be summarized as *Grow, Make, and Support Locally, Everywhere.*

We are now seeing powerful innovations originating in many countries. It is important for all countries to embrace the best innovations together rather than compete if we are to avoid a climate disaster. Many citizens of industrialized nations have a good quality of life, which reduces the number of citizens demanding change. However, to retain human sustainability, these citizens must use all human powers of innovation to raise the revenue to buy the products and services needed to meet the rising aspirations of all citizens. In finding ways to avoid a climate disaster, all nations must create sufficient well-paid jobs to replace jobs that become redundant.

China is increasing its prowess in converting emerging technology to make clean products for global markets and making the whole nation a desirable and sustainable place for all its citizens. This is strongly led by President Xi Jinping, who is a trained chemical engineer. China saw the need for electric transport and took steps to secure a world lead. This has been extended to embrace a hydrogen economy. There remains massive scope for world leadership in many products and services not yet covered by anyone at the necessary world market level. Many industrialized countries, including the UK, have allowed important skills needed for the future to decline, as employment has been exported to nations able to manufacture and provide services at a lower cost than the domestic equivalent. These countries must find the means to engage all their workforce productively.

Recovery from the COVID-19 pandemic, and the disruption of trading arising from the rise of increasingly divisive populist politics and policies, focuses on the urgent need to fix weaknesses in global mass markets while the need is visible for everyone to see, and also while interest rates are low. It has never been easier to address new markets across the globe, as we can speak with anyone wherever they are, using technology that has been with us for over a decade. Smart large companies are funding educational scholarships around the world, where the local national buying power is expected to grow. This strengthens bonds between countries, but it can result in political dominance that has dangers in terms of cultural freedom and reduced diversity.

Mother Nature has given humans the collective ability and guidance from the natural world to provide the means for an increasing world population and to live with improved lifestyles that are affordable out of earned incomes. But it comes at a cost, and one we must understand before we can manage it.

One of the most important principles in the foundation of understanding where best to focus our efforts in the war against climate disaster comes from Swedish physician and academic, Hans Rosling, whose 2018 book "Factfulness", based on work through his "Gapminder Foundation", has been instrumental in breaking free of decades-old misconceptions, which to this day, continue to misguide our understanding and approach to the key challenges faced by our global society.[4]

Rosling provides a framework to simply understand the changing behaviour of the global population by shunning the traditional and, in his view, woefully outdated misconception that the world is divided in two: the "developed" and the "developing" world, the poor and the rich, or even "the West and the rest". Rosling highlights in an accessible way that the bulk of Earth's population (c. 71%) is neither very rich nor very poor, but somewhere in between. He goes further to show that it is far more helpful to us, especially when considering the issue of climate change, to understand that the Earth's seven billion people are spread, Pareto-like, across four levels of income. From the poorest, who exist on less than US$2 per day, the next group (Level 2) who exist on up to US$8 per day, the next

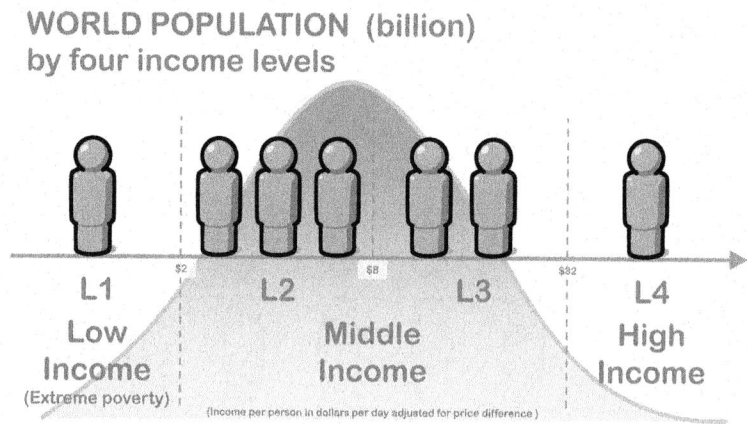

WORLD POPULATION (billion)
by four income levels

L1 — Low Income (Extreme poverty)
L2 / L3 — Middle Income
L4 — High Income

$2 $8 $32

(Income per person in dollars per day adjusted for price difference.)

Figure 1.1 World population by income levels.[5]

group (Level 3) who live on under US$32 per day and the richest (Level 4), who exist on more than US$32 per day (Figure 1.1).

This model is helpful as it shows the natural aspiration of all Earth's citizens to move from the left to the right not just in order to access greater wealth, but more importantly to achieve the greater life outcomes that come as a result of increasing wealth (e.g. longer life, lower infant mortality, better education).

If we understand the model and natural migration from left to right, we can then start to really understand, not just the impact on the environment, but the largest contributors.

Rosling's work does just that as he and the Gapminder team carefully show how 50% of the world's CO_2 emissions come from the one billion people in Level 4, 25% from the next billion, and in fact, it keeps halving as you move through the global population in descending order, ranked by income (Figure 1.2).

The world's population is 7.7 billion, rising to a likely peak of 11 billion in 2100.[7] Each of the Earth's richest 1 billion people contributes about 25% from four broad sources: food, accommodation, travel, and everything else, including education and public services and non-food retail.

Each year, the world output of CO_2 equivalent amounts to more than 35 billion tonnes and is rising faster despite all our best efforts to suppress the accelerating increases. China is the biggest polluter at c. 29%, followed by the United States at 15% and the UK contributes about 1%.[8] In 2019,

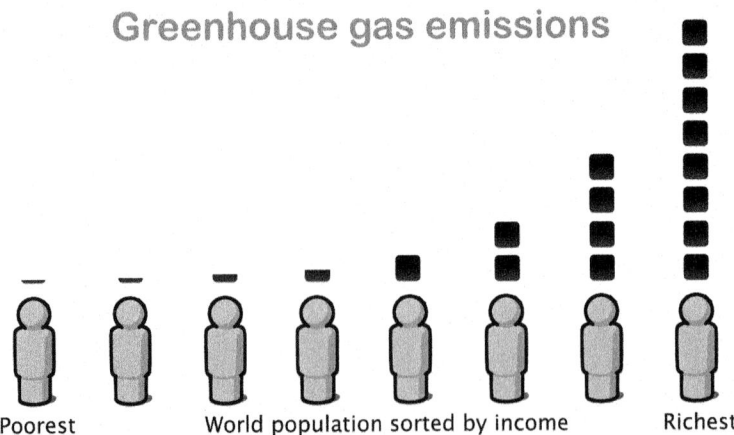

Figure 1.2 Greenhouse gas emissions by income.[6]

commercial aviation was about 2.2%[9] and shipping was about 2%[10] but both are set to grow. Commercial aviation may have been disrupted by the pandemic but is expected to see a return to a high growth rate to satisfy a huge pent-up demand. There is a real need for a proper solution in the shortest possible time, but we must get there from where we are now.

If we look at the worst offenders on a per person per year basis, what becomes clear is that looking beyond the outliers represented by the Persian Gulf states, it is the Level 4 nations (or more specifically, the lifestyles of the citizens that comprise them) that represent the biggest threat to the environment. This arises for many reasons, including spending power, which is often spent on goods with a high tariff in terms of CO_2.

Examples from Level 4 nations include the love of big thirsty 4 × 4 cars, foreign travel, and luxury cruises. But it is the more basic aspects of our lifestyle that present the bulk of the problem. One of the more significant examples lies with the diet of many Level 4 citizens, which is often rich in content produced in animals that produce methane, which as we'll see later in Chapter 11 is a very damaging greenhouse gas. Looking at the league table of meat consumption per capita, you can see that Hong Kong, the United States, and Australia eat far more meat per capita than most other nations (137 kg, 124 kg, 121 kg versus global national per capita average of 49 kg).[11] The implications to the climate from this kind of over-consumption only grow worse as an increasing number of nations join the Level 4 club.

As another more local example, many of the 27 million homes in the UK are old with serious leakage of heat. Similarly, much of the infrastructure is old leading to yet more CO_2. Many of the luxuries we surround ourselves with such as clothes, cars, and phones are changed too frequently to keep pace with fashion. In terms of their impact on the environment, UK citizens are twice as bad as the world average, while citizens of the United States are more than three times the world average. Many changes improve life quality, improve diet, and enable people and families to operate in a more self-reliant style of life, work, and leisure. Industrial nations have cut down their forests and removed other sources of CO_2 absorption by removing features such as peat bogs.

There are over 100 times more people on Earth who want to live the life of a mid-range Level 4 citizen, so Level 4 nations must change and do it quickly. The changes required are big, and many changes are necessary. It is crucial for engineers and the business community to show that we can take care of our planet while raising the quality of life for everyone. In many cases, we require the rich to pave the way, but we must move swiftly to the global mass market with all the solutions that have the potential to improve sustainable human life on Earth.

Clearly, as the wealth of people in the middle-income groups grows, so too will their consumption, which, in turn, exponentially increases their impact on the environment. As a result, we are quite simply now at the point (or beyond it) where we absolutely must replace fossil fuels with renewable energy. This significantly increases the amount of electricity we must generate because we will now have to replace all the energy previously supplied by fossil fuels used directly as petrol and diesel for transport, the natural gas for domestic and industrial process heating, and so on. Unfortunately, renewables cannot match our demand for energy 24/7 without energy storage and this, in turn, means we must find new ways of storing renewable energy in a way that can be delivered to the right place, in the right amounts, and in the right form, at the right time for the consumer. More than this, we must transition from where we are now, to a new and better normal, noting the huge inventory of equipment, with a long residual life held by the industrialized nations.

Our tendency for urban living in high-rise buildings, built using steel and concrete, involves the generation of large amounts of CO_2. High-rise buildings made from cross-laminated timber are now possible, with a far lower CO_2 penalty per unit of living space, without loss of quality. New rules are rightly tougher on the use of structural timber in high-rise buildings. As we will see later, there are other more efficient ways of achieving greener construction processes for residential buildings and business properties of high quality. When concrete and steel are used, we must move to processes that involve fewer emissions. Power and heating must be derived from renewable sources as the means of achieving intense heat and high power. More efficient structures that use less material are also important in keeping emissions down.

The point is, to avoid a climate disaster, it is important for people to reset the way they live, work, and play.

As with COVID-19, we need very prompt action on every front. While it would be marvellous to achieve this by consent, some enforcement may become crucial to save lives. Climate change is more serious than COVID-19! It is just less visible, and it is initially more gradual in its ruination of lives and livelihoods.

We need to look at what hurts most in terms of climate change leading to a climate disaster. We must also look at the actions that must be taken to prevent a runaway into unsustainable life on Earth. Many actions are required. Most are already known, but few are understood by those who must act. There is growing anger, but few plans are visible in the public domain. There are good signs that citizens are willing to make the changes, but there is an absence of information about what can be done and what should be done first to achieve the quickest and largest impact. Excellent work already done reveals the potentially powerful actions that must be taken. These people know that holding back will lead to a slide into oblivion. There has been a complete failure to get enough of those with the power to act to understand what specific actions to take. While some good plans are being implemented, climate damage is rising faster than remedial action. The evidence requiring ever stronger and faster action increases daily. There are solutions that offer strong benefits that must be rolled out on a mass-market scale globally to be effective and affordable. The provision of effective contributors must also make business sense. Most importantly, citizens need to feel comfortable with the actions taken that will impact the lives of everyone. Climate change mitigation needs the same urgency, right across the globe, as was the reaction to the COVID-19 pandemic. We cannot return to business as usual. We can, however, transition to a "new normal" that is better for everyone than the "old normal".

The change will be enormous, creating many valuable new jobs, but most of it will be people-led, as was the case with the Industrial Revolution, initiated in the north of England. This time, there is a greater need for global standards in our globally interactive world. There are more than enough resources to prevent a climate disaster, but the pace of implementation of powerful solutions is too slow.

Finally, all solutions must be safe in resisting all forms of adversity, such as the COVID-19 pandemic, plagues of locusts, and extremes of weather. Humans can devise effective means for controlling the problems, thereby minimizing the loss of life. There is no hiding place against a climate disaster, only a mass reset of all human behaviour will do.

Notes

1 The World Bank (2021), *Air Transport, Passengers Carried*. Available at https://data.worldbank.org/indicator/IS.AIR.PSGR (Accessed 3 March 2021).

2 The Royal Institution (2021), *Planet Earth: A User's Guide: Engine Earth – Lecture 1*. Available at https://www.rigb.org/christmas-lectures/watch/2020/planet-earth-a-users-guide/engine-earth (Accessed 3 March 2021).

3 MacKay, D. (2009), *Sustainable Energy –Without the Hot Air*. Cambridge, UIT Cambridge Ltd.

4 Rosling, H. (2018), *Factfulness*. London, Sceptre.

5 The Gapminder Foundation (2021), *Income Levels*. Available at https://www.gapminder.org/fw/income-levels/ (Accessed 3 March 2021).

6 The Gapminder Foundation (2021), *CO2 Emissions by Income*. Available at https://www.gapminder.org/topics/co2-emissions-on-different-income/ (Accessed 3 March 2021).

7 United Nations (2019), *Population*. Available at https://www.un.org/en/sections/issues-depth/population/ (Accessed 3 March 2021).

8 Global Carbon Project (2019), *Global Carbon Atlas*. Available at http://globalcarbonatlas.org/en/content/welcome-carbon-atlas (Accessed 27 February 2021).

9 The International Council on Clean Transportation (2020), *CO2 Emissions from Commercial Aviation: 2013, 2018 and 2019*. Available at https://theicct

.org/publications/co2-emissions-commercial-aviation-2020   (Accessed   7 March 2021).

10 International Energy Agency (2020), *International Shipping*. Available at https://www.iea.org/reports/international-shipping   (Accessed: 7   March 2021).

11 Our World in Data (2019), *Per Capita Meat Consumption*. Available at https://ourworldindata.org/meat-production#per-capita-meat-consump- tion (Accessed 7 March 2021).

2

WHY DESIGN ENGINEERING?

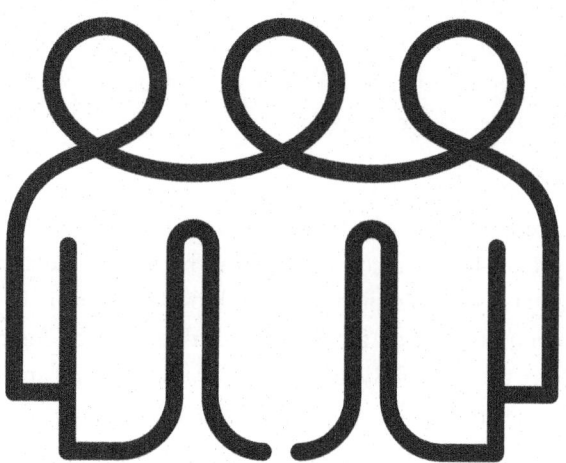

DOI: 10.4324/9781003193470-3

Over the course of my career, I have seen many examples that reveal why design engineers are essential in leading some of the most important paths that are needed to achieve a sustainable life.

Back in the 1970s, while thinking we had got our RB211 engine safely into airline service, we received news from airlines and passengers that our engines were backfiring with a loud bang and huge flames out of the back. Not exactly the passenger experience we had in mind when we designed them! Immediately, the best possible engineer with the most qualified experts quickly brought back the guilty engines for testing in a real effort to find the cause and provide a fix.

But on testing to try to reproduce the problem, the engine behaved perfectly, just as we had designed it and with no trace of any fault. I decided to dedicate my design engineering self to a critical examination of the overall design, asking "what does it take to make this engine behave so badly?"

Over a few long days, I found 13 little problems, where if just a few of them occurred together, the engine would cough in the same way that had been seen by the airlines and passengers. I started to think through how to fix every one of these possible guilty parts and found that one change could fix everything. However, because we could not repeat the backfire on a guilty engine, recovery action was delayed for further tests. At this point, I dropped by to see our manufacturing director. I explained what I had found and showed him my design, drawn free-hand on the parquet floor of my living room. He thought for a few minutes before saying, "That is easy! We have the material, and we can make the new parts quickly and get them built very quickly". He did just that. Our chief engineer said we would test it right away, and we would try extra hard to make the engine cough. The modification worked and was fitted to all engines. However, because we were all focused on a cure for the engine's cough, we failed to notice that every engine had become more efficient. The worth of the improvement was, in today's money, about US$1 million per aircraft per year.

This book tells this chief design engineer's view of the many pathways, created by exceptional people from all over the world, that are needed to improve human life in a sustainable way by taking the actions needed to alleviate adverse climate change and achieve cleaner air in busy urban areas. It is a story based upon fact and reasonable calculations plus forward projections over the next few years, using sound science and rules set by nature. Based on this, if such innovations are delivered and managed well, we know the Earth has sufficient renewable resources to support the resetting of the ways the world's growing population will live, work, and play.

Moreover, our planet is awash with funds making negative returns while awaiting a better investment opportunity. Collectively, humans have the knowledge to make a big difference. Powerful innovations are being demonstrated by people from all over the world. Preventing a climate disaster requires a global uptake of all the best innovations, with the consent of citizens, and in a way that makes business sense, sufficient to attract the investments needed.

Co-ordinating the necessary actions fast enough to avoid a climate disaster is the real need. This requires everyone across the world to see the need for fast action at a global mass-market scale. Complete compliance across all nations is essential if a total wipe-out of human life is to be avoided. There is also a need to find ways to cover the high cost of investment to take proven solutions from proof of concept to full-scale production at affordable prices in bankable businesses. The task is made more difficult by the interdependencies between the many developments, and the need for global standards.

As changes go into service, experience will highlight the need for further refinements, adding further to the complexity. Humans must also transition from where we are now, at a pace that is much faster than natural attrition. Advanced nations have huge amounts of old stock with a long economic life remaining. While recognizing change needs to happen, and at a mass scale, throwing away perfectly usable, serviceable infrastructure simply adds insult to injury and massively dilutes the benefit of any new "cleaner" replacements. Solutions must be found for cleaning up the ways we use these legacy assets until they can be removed from service everywhere in the world.

Advanced nations rightly treasure their human rights, and their freedom of action, but sometimes this can undermine necessary compliance in terms of living and working in better ways that also address climate change. We have all witnessed the almost unfathomable power of misinformation and conspiracy theories being "weaponized" to subvert societal behaviour. Watching scenes of rebellion against mask-wearing, social distancing measures, or vaccinations has been one of the more bizarre spectacles of recent years, especially when viewed from a scientific and/or engineering perspective. But the paradox of human rights has never been more clear or more fundamental to our survival. Do our human rights give us a licence to kill or cure society? Should we allow freedom of expression, or should we force individuals to adhere to specific behaviours? These are tricky questions to answer, and while I appreciate there are clearly more issues involved than simply freedom of expression alone, it is clear to me that we can no longer afford to evade these issues within our national and global dialogues.

The COVID-19 pandemic has provided such an example in the short term, but climate change presents a much more devastating problem over the long term. The pandemic showed how more disciplined nations, such as New Zealand and Singapore, were able to control the virus much more effectively than other nations, such as the United States and the UK, who sacrificed such results in return for greater freedom of action and expression. The cost of the difference in approach can be counted in the difference in COVID-19 deaths per capita. At the time of writing (4 March 2021), COVID-19 deaths per capita for New Zealand and Singapore are about five per million, whereas, for the UK and the United States, they are 1,865 per million and 1,584 per million, respectively.[1] While that comparison may feel cold, crass, and brutal, it is but a tiny taste of what might be to come when one recognizes that the total impact of COVID-19 is but a fraction of the potential impact on human life that will be caused by climate change.

A climate disaster is a far greater threat to human life than coronavirus. Avoiding it cannot be achieved by lockdowns or vaccinations. Instead, our inoculation from climate disaster needs action from every human on the planet and requires them to adhere to rules that will improve the quality of human life for everyone. For the necessary universal compliance, the reward must be good for everyone, but even then, there will be a severe challenge in getting everyone to accept the need for new ways of living, working, and playing. However, I am convinced with the right innovations, supported by the right stakeholders across our national and global societies, achieving such a change is still a realistic and achievable goal.

Building new assets incurs emissions penalties, so we must find the means to achieve everything we want, for the lowest level of greenhouse gases. The first question is, "Do we need this?" If we do, can we use less material? Can we find/make the materials with much lower emissions? The electrification of many assets, such as cars, incurs a bigger emission penalty than the petrol-powered original, forcing the question, "How do I improve on the basic model with much reduced emissions?" Electric cars with high performance and a long range between charges need very heavy batteries, giving rise to higher climate change penalties. Can we make better batteries, and cars that need less battery capacity? Buildings, roads, trains, and other such components of our built environment all incur penalties, so it is important to use new and cleaner methods in their making. The new way of living and working impacts the design of almost every asset. This increases the incentive to employ more efficient and cleaner processes of designing and building, and to operate over a longer life. Provision for updates should be included, where possible. Necessary changes cover where we live, what we eat, how and where we work, our sports, travel, and entertainment. It is vital for every one of these changes to make things better, while at the same time preventing a climate disaster.

Complete lockdown in many countries, in reaction to the coronavirus pandemic, proved such a broad, deep change was possible and all in a relatively short space of time. Disaster from climate change represents a far bigger threat, but somehow it is less visible, even though extreme adverse weather events are increasing in number, geographical reach, and severity. Moreover, it seems likely to remain that way until the disaster is upon us. This results in our current climate-damaging activities continuing unchecked.

Just as people wear masks to save others, and themselves, from the dangerous virus, we must all realize the role of every human threatening the climate we all share. To this end, we must engage with the young by addressing their anger about their future being put at risk by the damaging behaviour of earlier generations. They have skills with social media that give them instant global access. We must use these media to change behaviour, while just a few years remain to reset all human behaviour in order to avoid a climate disaster. But our approach needs to be centred on extolling the path to a better life for all that is sustainable, rather than an excessive emphasis on impending climate disaster. In order to have maximum effect, we need to inspire people into action, not frighten and threaten them.

The reset depends upon the proper use of new technology by everyone including poor and disadvantaged citizens. Instead of spending huge sums on infrastructure designed for the outgoing unsustainable ways of living, it would serve humans better to provide every child with access to good technology, ideally built where the children live, to world-class standards. This may need licence agreements and royalty payments, but every country needs jobs. It is more than a double win.

If jobs are created to build the best technology for everyone, this then enables humans to increase the number of people able to gain well-paid jobs requiring digital skills that, in turn, will gain more jobs in a rising world market, while increasing the scope for more support for families, thus decreasing the demand on services provided by the state. Social media has the potential to give everyone insight into best practices and this will increase competition for a market share of rising global investments and markets. The risk here is that if you are not with the front runners, there will be insufficient funds to buy from competitors, and nations who miss out risk being left behind.

This will present a tricky balance that will need to be struck between national and global interests. Heavy hitters in invention and innovation such as the UK, the United States, and China need to use the best technology from wherever it originates to form global partnerships and avoid debilitating excess competition. While some competition is a valuable market force, too much will not only stunt growth and adoption, it risks de-railing any potential overall progress. High investments are needed to get products right for the best price. If there are too many competitors, then few if any can raise sufficient funds to achieve the best product at the best price. Every participant becomes a loser!

So why am I convinced that design engineering provides the answer to solving our climate problem? Well, achieving sustainable life for all requires many innovations, all of which are interlinked and carry some risk. First, the design analysis must show the absence of all dangerous side effects for each change. Tests by an independent group must deliver proof that the design is safe and makes the outcome better. The aeroengine I mentioned at the beginning of this chapter, with its tens of thousands of parts, each dependent on the rest, is comparable to the many changes people must make if our lives on Earth are to be sustainable. Based on my experience in aviation, to make life better, design engineering thoroughness is the only way for safely delivering complex interactive changes.

Note

1 John Hopkins University (2021), *Mortality Analyses.* Available at https://coronavirus.jhu.edu/data/mortality (Accessed 4 March 2021).

3

HOW DID WE GET HERE?

DOI: 10.4324/9781003193470-4

My parents were from one of the first generations to really benefit from the full systemic impact of the Industrial Revolution. My mother, Eva, was born in Leigh, Greater Manchester, in 1906. Her elder sister persuaded her to work hard at school and she won a place at the local grammar school. She got a job with an industrial company in Manchester requiring her to commute daily by steam-powered trains. More than this, she often commuted to take night-school classes in Bolton where she gained certificates proving her increased skills, enabling her to get a better-paid job. I still have her certificates, dated 1922, from when she was 16 years old and whenever I look at them, it makes me truly connect with the scale of the kind of societal benefit that engineers such as Watt, Stephenson, and their peers enabled. None of what my mother (or indeed, as a result, I) was able to accomplish would have been possible without the incredible innovations they brought.

The legacy of the Industrial Revolution

The Industrial Revolution began in the UK with the invention of a practical steam engine by James Watt in 1769. The first demonstration of a steam-powered train took place in 1829, with Stephenson's Rocket demonstrating that the steam engine could be carried on a train and have enough power to pull carriages at a reasonable speed with a full load of passengers. This first passenger railway ran between Liverpool and Manchester in the UK.

In the years that followed the system expanded, covering an increased number of routes, such that by the early 1900s, many passengers could be carried for a price that anyone in full-time work could readily afford. It could also transport heavy goods economically. More than this, it could take those who wanted work to their place of work in giant factories and mills where goods, used by everyone, could be made and sold for affordable prices. These prices were competitive in global markets, and the UK became the "workshop of the world", from which many UK citizens shared in the gain. Just as importantly, it allowed UK citizens to travel for education and training leading to better-paid jobs creating more and more value.

The Industrial Revolution gave new jobs in mining coal and metals, and new infrastructure. There were new jobs in making goods, textiles, and clothing. Shipbuilding expanded greatly. Mass production in huge new factories greatly reduced prices, with consistently high quality, giving access to much-increased markets. Domestic markets expanded as a direct consequence of more affordable products, previously enjoyed by the few. The mass-produced articles soon overtook the earlier handmade products in terms of performance, reliability, and price. Year by year, products got better as users indicated scope for practical improvements in design and manufacturing.

This allowed markets to grow, together with increased employment. In turn, this led to more consumption by more people and introduced new possibilities that were previously unavailable such as people being able to take holidays away from home.

However, alongside the improvements in wealth, health, and lifestyle, and propelling the world into the growing prosperity of a new industrial era, the UK had, without realizing it, initiated "Climate Change".

It took another 80 years from the start of the Industrial Revolution in the UK for the rest of the world to catch on to the huge gains from industrialization, and full employment, with some countries with much bigger populations taking longer still. The countries that followed the UK also failed to realize the damage being done to the Earth's climate. It is the global expansion of industrialization that is the source of the problem, made worse by people wanting to heat their homes, and generally indulging in things that make life better, without realizing the added damage to the climate. More than that, there are at least 100 times more people wanting to live a life at or above that of the average UK citizen, but the UK way of life is not sustainable.

How climate change escalated

Records show that in the 1800s, the population of the UK grew rapidly. The Industrial Revolution played an important part in that population growth through improvements in food, clean water, and a higher life expectancy. By the year 1900, the UK population had grown to an estimated 40 million, which represented 2.3% of the estimated 1.7 billion population of the world. The British Empire was estimated to be 384 million, second only to China, and much higher than the United States. The UK's population has grown two-thirds larger since 1900, while the world's population has grown by 4.5 times. More than this, spending power in the industrialized nations has significantly increased, with much of it spent on food, products, transport, and services that carry a significant greenhouse gas penalty.

The gains came with a high cost in terms of climate change. The gains were enjoyed, but the climate change penalty was not observed until the hurt started to become noticeable. Scientific observers drew attention to this, but too many leaders failed to recognize the importance of climate change, and the increasing prospect of a climate disaster.

The role of humans in the adverse change was disputed by many. It has taken about 170 years for scientists to find the data that enables people to accept that humans cause climate change and that everyone needs to contribute to the massive collective efforts needed to prevent a climate disaster that will take place within the lifetime of many people living now unless preventative action is taken without delay. There have been many earlier warnings by experts, but these have not been acted upon, resulting in a need for widespread action to take place with real urgency. Most people assumed that their contribution to atmospheric pollution was too small to hurt the climate. The atmosphere was assumed to have an infinite capacity to absorb all the gases and dust being dumped into it.

Despite the growing amounts of evidence, there are still large numbers of citizens who deny climate change. For everyone, it is important that the changes we make are good for individuals and national well-being. The changes must also have the potential to make sound business sense.

Lessons learned from the Industrial Revolution

The Industrial Revolution provides many lessons, some of which appear to have been ignored by the UK, in the current technological revolution. In the Industrial Revolution, the ordinary citizen saw great advances in life quality, health, education, and training, all arising from British invention, innovation, and a willing uptake by more workers. Progress was slow at first, but by 1920, the gains were widespread.

It is important to remember that the Industrial Revolution started in the north of England, in Lancashire, Derbyshire, and the northeast. Men of vision saw a way of using cheap power to employ many workers to make goods of quality, at prices the workers could afford. Water and steam, raised by burning coal, powered industrial processes that made mass production possible when supported by human labour, and sometimes even child labour. The work for humans was hard but it gave families a better lifestyle than previously was the norm.

In the current technological revolution, countries like the UK, the United States, and others remain as some of the leading nations for innovation, but all too often, employment emanating from this innovation goes to countries with bigger markets and higher potential growth rates. This business model is good for multinational companies, but how this is used to benefit local citizens needs further thought to retain and grow support services provided by the state. If we fail to bring benefits to citizens locally, we won't be able to get them to implement the changes that are going to be required in order to solve the problem globally. Avoiding a climate change disaster requires a drastic reduction in personal travel and the transportation of food and goods until we can reduce emissions to net-zero. This favours a return to increased self-reliance at the national scale. This is also true for every industrialized nation.

The coronavirus pandemic has shown the fragility of dependence on component supply lines from many countries. A lockdown in any one country can destroy businesses and put many people out of work. Tensions between nations add to the risk in supply lines. Each country needs sufficiently well-paid jobs to be sure of a sustainable life for everyone. Exporting jobs may give lower prices but if the citizens of the buying nation have no work, citizens cannot buy. When much of the value-added results from better technology and better tools, the importance of low labour costs reduces.

With the replacement of old jobs by new ones, there is an opportunity to grow and make more products locally. New technology allows increased productivity, and the opportunity to create more new jobs locally. New products are emerging where any nation could be a leading producer for world markets. Individual nations such as the UK should strive to achieve this major source of national income in a way that benefits every local citizen while delivering global benefit. Increasingly, more trading will be done electronically instead of the mass movement of goods and people. We must think more about how we carry out international trade, using investment to build, make, grow, and sell locally, with returns on investment being paid electronically to the investors and owners of the knowledge and experience used.

This is different from make and grow local but sell global where the best natural provisions prevail, leading to excessive travel costs and emissions, causing climate damage, and even creating global pathways for virus transmission. Growing and making locally changes international trade, but it does not destroy it. We do need technology and innovation from overseas, as well as inward investment. Wherever possible, money and data must do the bulk of the moving, not people and goods. However, successful investment and trading depend on mutual trust, and a good understanding of the people and cultures involved, while modern digital and electronic forms of communication are good, they serve only to reduce the frequency that face-to-face interaction is required – they do not replace it.

We will still need to travel, and this includes air travel over long distances, but even with significantly reduced frequency, travel and transport must reduce their emissions. Zero emissions are the requirement, but we must get there using what we have, using cleaner fuels until we have the transport systems with zero emissions.

For individual nations to succeed in global markets, national businesses must use the best available technology, from wherever it originates. In some cases, good nationally originated technology has been improved upon by other countries in ways that raise performance and lower cost, resulting in increased market share and a wider range of applications.

In earlier times, improvements to products and services were initiated by customers saying, "Could you do this for me?" In more recent times, a company with great technology or experience might see your good first model, and say, "Wow! That gives me something I can improve on". The approach of finding the best and making it better gives market leadership.

As with all new technology, we need to exploit its full potential. Too often we see powerful new technology used to speed up the old, and unproductive, processes, rather than using the technology to lead to a much more productive process, with higher margins and increased market share. If it was "efficiency" that formed the heart of the Industrial Revolution, it is "effectiveness" that should do the same for the emergent technological revolution. Pitting humans against machines in the workplace leads to a race to the bottom as the diminishing return of the quest to efficiency costs jobs and ultimately cripples economies and all without significantly changing the outcome for the climate. Conversely, combining the best of machines with the best of humans provides an opportunity that is far greater than the sum of its parts, enabling greater outcomes for individuals, economies, and the environment.

There are additional ways of gaining global market share from the manufacture of licensed technology from innovations created overseas. My experience of trading with big countries aiming to become world-class players highlights the technique used by most industrialized nations.

During my time as UK adviser to the government of the Republic of Indonesia, I watched this play out time and time again as competitors of the UK from the United States, Japan, Germany, and France were observed all using similar procedures. As the UK, and its competitors, try to sell products and services, the foreign buyer says: "You provide the money and give me your technology and know-how, so we can make some of the high technology parts. You will be paid out of the success of our using your product". There are real dangers with this business model, as Rolls-Royce has found to its cost. If the buyer does not use what he has bought, then the owner of the know-how and the investors no longer get their return on their costs and investment. It all needs thinking through, from the point of view of risk and revenue sharing, within a global context.

The need for action on a global, mass-market scale

Every day, as I write this book, I am made aware of better ways to live and work, with good ideas originating from all over the world.

Experience tells us that problems this big must be broken down into bite-sized bits that can be managed effectively in a way that must meet with the approval of the public and the investors. In terms of CO_2 emissions, the Persian Gulf States are the worst offenders on average per person, at around ten times the world's average.[1] As we will see, citizens of the countries that occupy Level 4 in Gapminder Foundation's World Population by income model (see Chapter 1) are the worst offenders, on a per person per year basis. Citizens of these countries are proud of their lifestyle, without realizing that it has a high dependency on some of the worst offences from a climate damage point of view. This includes power from coal and fossil fuels, a diet with a high dependency on the high consumption of meat, and income from foreign sales of harmful products. Currently, governments in Level 4 nations seem increasingly committed to delivering a leading pathway to responsible global action. There are good signs. These countries have the skills, resources, and domestic market to move fast. Several of the Persian Gulf states are also well advanced in this transition. We can expect to see competition in the global energy market for the cheapest clean fuels for transport, heating, and power, leading to products like steel and concrete with much lower emissions.

Action at a global, mass-market scale is vital. Without this scale of implementation, the changes are not affordable, and the benefits in terms of avoiding a climate disaster are totally ineffective. Much innovation will take place with minimal government intervention, but there are vital elements where governments must take a stronger lead. Researchers must be well funded, and there is a vital need to watch, then embrace, the best technology from wherever it originates. There is excellent technology and practices emanating from the United States, Canada, France, Germany, and many other places. From meetings I have attended in the UK, I sense too much inward-looking to our own capability, often at the expense of better ideas that were generated elsewhere. It appears that there is also a failure to recognize the need for a total systems approach. Climate change is a global issue and universal standards are going to be required across the globe, and governments must work together, and with businesses, to get full interoperability over all fields of development.

For many industries, the scope for improvement is far greater than people expect, sometimes in excess of 100 times. Most important of all, we must take a total system approach since every change impacts every other change. Happily, we have powers of near-instant communication, meaning that we can consult experts across the world, wherever they live.

Every country must find a way for its citizens to share in the aggregate wealth created from their innovations, as well as wealth from using products and services, dependent upon innovations from world sources. All countries will want their innovations to reach global markets in a way that benefits all their national citizens.

China, India, and many of the countries in Southeast Asia have huge pools of well-educated people, with training in science that is suitable for generating new technology, and the creativity to design, make, and sell new high-value products and services that are affordable by huge populations around the world.

Many nations are discovering that they have massive resources that had no value in the world we are leaving, but now have exceptional value in the world we must move into to achieve sustainable human life. The Sahara Desert has more than enough solar energy to meet the present world demand for electricity. Until this resource is harnessed, the desert can support very few human lives. Many of these communities live in poverty, with disease and little help in improving living conditions. We will show how that can change. Many countries have resources that had low value in the lifestyle we must leave. In the world we are moving to, many of those resources become valuable, with the potential to improve lives locally. Examples include solar power from Australia and the Mediterranean countries. The United States has considerable natural resources that until now have had little value. Oil and cattle-rich Texas receives generous gifts of solar radiation. Such resources will help in the reset from fossil fuel to cheap solar electric power with clean and easy-to-use stored energy.

Nature has allowed humans to learn and use all the tools inspired by nature to provide all that is needed to satisfy the increasing demands of the world's growing population without increased damage to the global climate. The first humans depended almost entirely on gifts from nature. As populations increase, better tools and technology are needed. Increased threats from disease and other forms of adversity increase, adding the need for even more powerful tools. Scientists lead the search for the new tools, while engineers design, build, and use them, and the public sector, investors, and bankers provide the funds, leaving designers to refine the products in terms of their elegance.

The engineers must ensure safety, making sure that life-threatening risks do not take lives. Moreover, engineers must carry the public with them. Change involves risk. If adversity can happen, it will, so people need to know that the designer introducing the change has the power, and resources, to recover from any potential dangers.

As the chief designer of one of the world's most advanced aeroengines powering wide-body jets, there were some severe setbacks, despite world-leading thoroughness in design, manufacture, and operation. Much of my time was spent with the bosses of airlines, showing them how we were keeping passengers safe, when adversity strikes, without the airlines losing revenue. The transition from where humans are now to a more sustainable way of life will expose unforeseen problems, despite every best effort to avoid them. Human life must carry on safely, while the problems are fixed, enabling a return to the best possible performance.

The most onerous tasks include that of getting sufficient support from the people, especially in strong democratic countries, and the financing of the expensive transition from demonstrated solution to mass-market roll-out in a way that satisfies all relevant parties. Clearly, our governments and business leaders must be convinced about change for good, but we must take people with us, including citizens too young to vote. Early visibility through social media will be vital. Once teenagers see the gain, positive support will follow. Moreover, critical comment will help improve the new developments and accelerate the uptake. No generation has yet had such a powerful tool. Social media can play a pivotal role in motivating everyone to reset the way they live, work, and play before time runs out to avoid a climate disaster.

Note

1 Global Carbon Project (2019), *Global Carbon Atlas.* Available at http://globalcarbonatlas.org/en/content/welcome-carbon-atlas (Accessed 27 February 2021).

4

THE TECHNOLOGY REVOLUTION

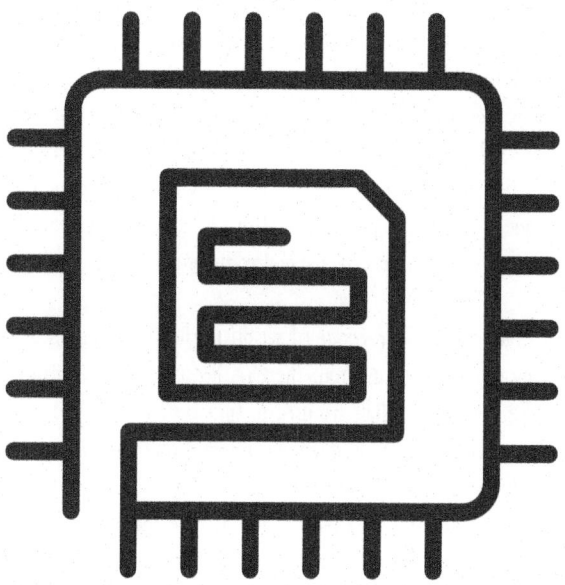

DOI: 10.4324/9781003193470-5

Half a century ago, as chief designer at Rolls-Royce, my colleagues and I experienced two commercial flight events caused by engine failures that could have resulted in a serious loss of life, and which threw a suspicion that many aircraft were at risk, including those with engines from competitors.

All the experts from all the companies that might be exposed came together to identify the problem. Instead of grounding every aircraft, we had enough reliable data to enable the company to remove the troublesome part before it became close to failure, such that we could safely fly, thereby keeping the revenue flowing, albeit with a slight increase in cost.

I remember it well. It was Christmas, and we were home with family when my colleagues and I each got a call. There had been a severe engine failure, over Chesapeake Bay, but we did not yet know what the cause of the failure was. Thankfully, everyone was safe, but this really got us worried, and so a crack team was despatched to get all the facts.

Then, within a day or so there had been a second failure, which sounded similar, but this time with another airline over Albuquerque. Again, thankfully everyone was safe.

Within hours of that first failure, we knew that our fan disc had burst at a life of only 1% of its design safe-life. Our fan disc carried 33 blades each trying to pull the disc apart as a result of their rotation, with a force of 75 tonnes. All big fan engines had a titanium fan disc like this, so did this mean every wide-body jet was in trouble?

In matters of safety, all parties come together pooling their expertise until it is certain that we know who owns the problem, and the responsibility for the fix. Within days it was clear that while we had been the most thorough, we had used our knowledge to take one step too far. Early one Sunday morning, we were to gather the team together to plan a safe way forward. It had snowed that night and the snow was deep and I lived at the top of a steep hill. I had cleared the snow, but black ice formed. There was no way for me to drive to the meeting, so I called my colleague with his old 4 × 4 Land Rover to meet me at the bottom of my drive. As I tried to descend the steep hill I fell and slid to the bottom on my backside, losing the heel of my shoe in the process. We got to the meeting in time and much to the amusement of my colleagues, I limped into the meeting, with trousers soaked and commenced to work with them on finding a solution.

A little later, a small team of experts and I set off to the United States to find out what we had missed. With us was a great guy from the CIA because we believed we were being spied on throughout the trip. For light relief, we used to spin the eaves-dropping spy long stories to raise a laugh. Once we were together, we quickly confirmed that our variety of titanium, which had a capability for high strength at high temperatures, was vulnerable to the manufacturing process needed to give the material its strength. From there the path to a permanent fix was clear and implemented at great speed with total thoroughness. The overall lesson is if things can go wrong, they will, so you must quickly find the safe way to keep the revenue coming in, while the business carries on with no loss of safety until there is a safe fix that returns us to full performance safely. That way you retain the confidence of your customer.

In this case, the critical part, the fan disc that carried the blades, was removed and replaced with a new one of the same design, to be used within its proven safe-life. This was repeated several times for every aircraft powered by our engines, until a new design of the guilty part, in a different material, involving a different manufacturing procedure, could be thoroughly tested, and proved to be safe for the longer design life, as originally expected.

The original design had a design life, supported by tests, of 30,000 flights, but despite this evidence, real-life operation showed the there was a previously unknown mode of failure that required the fan disc to be restricted to only 150 flights. We were confident that every disc made of titanium was safe for that reduced number of flights and so an expert team was despatched to change every fan disc at 150 flights. This enabled the airlines to continue with their service, albeit there were some costs, but these were well short of the loss of revenue.

The initial decision was taken within a few days based on actual test data. No failures took place. No one was alarmed, and because the revenue was maintained, there were funds and resources to cover the cost of the change. The faulty part was originally designed with exceptional thoroughness, but for a combination of reasons, the material, the design, and the manufacturing process delivered a surprise no one had suspected.

The warning from this experience is that every change we make must be thoroughly thought through to minimize the risk. Every nation should have plans to address the worst possible adversity because, if the worst life-threatening adversity can happen, you must assume it will. Every nation needs plans that consider the worst-case scenario and provide a way forward that enables life to continue safely, without loss of human life.

Modern technology allows us to minimize, or eliminate, the cost penalty for manufacturing locally, anywhere in the world. Climate change pushes nations to become more self-reliant to avoid a climate disaster. Making, growing, and selling close to the point of consumption also helps to avoid losses and waste associated with long transportation times. As populations grow, and pockets of adversity due to climate change increase, the need for this approach increases. For example, as natural disasters descend, be it in the form of earthquakes, forest fires, or plagues of locusts, a more robust way of feeding the people living close by is required for sustainable life.

"Make and grow locally, globally" is a useful description of the transition humans need to make. Put another way, let the data and funds do the travelling, not the product or the service support. Payment for data and service rewards the provider.

We must focus our minds on the importance of a local supply of all necessities. As noted in Chapter 3, the state needs to find the right way of benefitting local citizens for goods and services rendered overseas, using value-adding inputs from the local economy. This principle applies universally to all nations.

The coronavirus pandemic forced the need for global companies to buy land overseas and build new factories on that land with all the approvals needed. Moreover, new supply lines were needed for raw materials and packaging, and for shipping the finished product. Often, fast changes to plans were needed, as different countries varied their lockdown procedures. All this was done remotely as the managers responsible were not allowed to travel. It was done thoroughly, achieving good results, despite being managed and controlled, using technology, from employees working in makeshift home offices in bedrooms or spare rooms.

Another example that I've been exposed to shows a major UK company with factories in the Hubei province in China, and elsewhere in Southeast Asia, supplying huge volumes of antiseptic liquids and wipes to countries all over the world to help contain the deadly virus. The daily detailed management is controlled by the production director, our eldest son, from his apartment in Singapore. The board in the UK is in constant touch with all the necessary information to deal with the surprises that any emergency throws up. The UK company has, rightly, banned all travel for as long as the coronavirus pandemic persists. The whole operation is digitally controlled, with complete visibility of the facts, from which the necessary good decisions can be taken. This way of working has potential problems, associated with the links in the remote-control train. Adversity hits the different links in different ways, leading to a break in the chain that restricts output, even for the most important life-saving products. In normal times, the process controller would get the right experts to go and fix the broken link, but restrictions on travel, society lockdown, and other issues mean that other ways must be found very quickly. Increased self-reliance at a national level reduces the need for travel and the transport of goods.

Recent experience with lockdowns needed to limit the spread of coronavirus has shown that we can get help in digital form, to places needing it, far quicker than by personal travel. Both parties need to have the right equipment and the digital skills to use it. My wife and I see it first-hand as we use it to stay connected with our family, which is spread out over thousands of miles. Thanks to its use, we can see each other more frequently and more efficiently than could be achieved by travel, even for those family members that live nearby. Moreover, it keeps the family bond, illustrated by many actions of mutual support.

The message is clear, digital technology can reduce the need for travelling and the shipment of goods. To be clear, it does not eliminate the need for travelling but it does mean we can drastically reduce the frequency, providing more available time for activities that add more value to our endeavours.

In earlier examples such as the SARS virus, digital technology did much to save the day, but once that virus was contained, everyone reverted to exactly how they worked before the virus. This time, we must learn the lesson. Avoiding a climate disaster requires a reset to the way humans live, within which there is a high dependency on digital technology, in perpetuity.

The technology revolution, taking place now, has some important differences from the Industrial Revolution, in that technology is more readily available across the globe through faster and more effective means of communication.

For the last few years, a colleague and I have been monitoring a very wide range of innovations from across the world and thinking through how these help people to tackle climate change, air quality, the way we live, the way we travel, the way we take care of each other, and more. What becomes clear is the rate of increasing innovation is accelerating fast, in many countries. The accelerating pace of relevant innovation affects everyone, and it increases the number of new jobs that were not possible until now. This work also reveals the dying industries, and the massive infrastructure investments that may not serve the future well, such as HS2 (a high-speed railway route connecting the north and south of the UK) and Hinckley Point Nuclear Power Station (a 3200 MWe nuclear power station currently being built in the south west of England).

In February 2020, the UK government took the decision to commit to HS2. The railway hopes to encourage huge numbers of people to travel between the north and the south of the UK, in both directions, throughout the day, and every day. Indeed, unless this happens, HS2 risks become a giant white elephant, a shining metal shrine dedicated to a way of life that will have long since departed.

Climate change, the avoidance of virus spread, and global competitiveness (based on "fibre to the home" broadband penetration where the UK ranks #31 out of 36 countries)[1] say clearly that this project is taking the UK in the wrong direction. Even without considering the significant CO_2 penalty that will come with building HS2, the project represents a Victorian solution to a 21st-century problem. In Stephenson's world, the old way of working, data movement needed people to move while carrying the data on paper with them, often in a rather posh briefcase. Now data can be sent, and the action started within seconds, anywhere in the world. If Stephenson were alive today, he wouldn't be building a faster rocket, he'd be figuring out how to lay fine glass cables to every home in the country, taking an engineer's approach to outthink the problem with a solution that can be delivered to 100 times as many people, in half the time and at a fraction of the cost.

Shenzhen in southern China, close to Hong Kong, but on mainland China, is a rising example of extremely rapid progress. Indeed, many observers see Shenzhen replacing Hong Kong in terms of wealth generation. In around 40 years, Shenzhen went from a small border town of 30,000 people to a vibrant city metropolis of over 12 million people. Here, in what has rapidly become one of the world's greatest hi-tech megalopolises, creative engineers and businesspeople are using technology to create new and desirable products that people all over the world will want to buy.

Just as in my example of the problem of the fan disc, in every life-threatening crisis there is a need to avoid the threat while preserving the ability to recover to a new safer way of living, should adversity strike. While brilliant strides have been made by central governments, more needs to be done by working with businesses to find safe ways of keeping more of the economy running to preserve the fastest way to a sustainable recovery. From my own experience of tackling very tough problems, I cannot overstress the importance of mutual respect in reaching the right course of action, with real substance that enables everyone to understand, and pull together, to complete the agreed actions.

Replacing fossil fuels with renewables plus energy storage, and new ways of moving stored energy represents a gargantuan change to a seemingly infinite number of systems around the world. A change of this magnitude leaves considerable scope for unseen and hidden surprises to emerge.

Plans are needed, at the start of such a change, to identify, then prepare a safe recovery, should any of the possible "worst fears" occur. Doing this requires advanced modelling techniques, and the technology to do this exists now.

Humans must be ready with dependable plans to continue to operate safely, until such time as there is a proven fix for the surprise occurrence of adversity. The technology revolution means that collectively we have the technology and tools we need for the safe approach, but we must not let tighter control over change stifle innovation, because technology-led innovation is essential for sustainable human life as the world's population expands.

Note

1 OECD (2021), *Broadband Portal – Fixed and Mobile Broadband Subscriptions per 100 Inhabitants*. Available at http://www.oecd.org/sti/broadband/broadband-statistics/ (Accessed 5 March 2021).

Part 2

CLEAN ENERGY

Critical actions

- Maximize renewable power
- Move to a global green hydrogen-based economy
- Maximize every resource (natural and human-made) capable of absorbing CO_2
- Move industries to renewables plus energy storage
- Make nuclear power smaller and more affordable

DOI: 10.4324/9781003193470-6

5

ENERGY STORAGE SYSTEMS

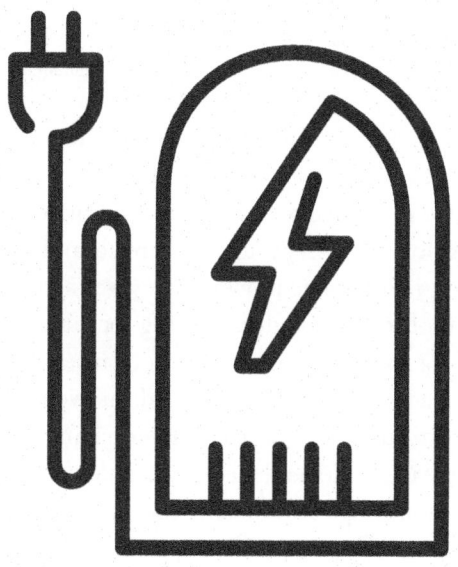

DOI: 10.4324/9781003193470-7

As a schoolboy, I was a regular visitor to the Farnborough Airshow, where I admired the first passenger aircraft with sleek lines powered by four jet engines hidden away in the wing roots. The de Havilland Comet entered passenger service in 1952. This giant step attracted attention all over the world. As I started my career with Rolls-Royce, I got the chance to work with Geoff Wilde, a highly innovative engineer leading the company's advanced projects team. I recall a top Boeing advanced projects design engineer visiting and saying that the best place for the engines was in pods under the wing for easy access, and for the subsequent fitment of more efficient jet engines to follow in the years to come. The Boeing 707 entered passenger service in October 1958, powered by four jet engines in pods under the wing. This is the way we do things now, except with engines so reliable they can stay on the wing for about a decade with only routine servicing, backed up by continuous surveillance, we only need two. This illustrates how the first step started in the UK but the United States improved a good idea making it better and more practical. There were other lessons. From this, I learnt that when designing something new, make sure you know how it will be used. Designing the engine in my office in Derby is a long way from when the engine is in operation in places with extreme climates. The designer's cold winter's day in Derby is a long way from the maintenance engineer standing on the tarmac in Winnipeg in winter when it's −40°C, with a heavy blizzard trying to check the oil before we fly. He's wearing thick gloves, it's dark, and the engine is still hot. As the designer, I must know the maintenance guy will get it right every time. As we start to look at new ways of storing and transporting energy, for example, when running a zero-emission aircraft running on fuel at −253°C in an aircraft of a radically different shape (we'll hear more about this later), every detail must be safe under all circumstances. We need to be open to thinking about both known and unknown scenarios. We need open, pragmatic, and logical thinking to help us minimize the risks of unforeseen challenges and dangers.

Nations must change to renewable power, but that is only possible when backed by sufficient energy storage, most of which must be transportable in the right form for mass-market use at the right price.

More and more of the energy used to create and provide goods and services must change to electric power. This gives a clear benefit in terms of reducing greenhouse gases and improving local air quality at the point of use, and this is of extreme importance in terms of health for those living with or near places where power is used. As we discussed in Chapter 1, as the use of fossil fuels is eliminated, the demand for electrical power will massively increase. A dependence upon renewable sources of energy is not possible without huge amounts of energy storage. Energy storage is only viable when it is delivered in the right quantities, at the right time, place, and price. Of the many forms of energy storage, just three will likely dominate: batteries, sustainable liquid fuel, and hydrogen.

Batteries

Batteries are strong contenders for a fast response in static industrial, domestic, and some short-range transport applications. Tesla is a leading company with many rivals, especially those trying to improve affordability by mass markets because without global mass-market uptake batteries will not be effective enough in avoiding a climate disaster. We can expect further developments in battery-stored energy per unit weight and in terms of recharge time but increased charging rates might seriously hurt battery life and hence cost. While we can use batteries to power most things, we do need to remember that as we remove fossil fuels with energy supplies needing renewable electric power, the demand for renewable energy rises steeply. We can moderate that by choosing to downsize the demand. While present-day lithium batteries do an excellent job, the cost is too high to meet mass markets globally, and as such, battery-powered cars just do not do enough good for the climate. Every electric car helps, but there needs to be a big reduction in price and in the practicality of use to get the numbers needed to make enough of a difference.

Researchers at Penn State University in the United States have found some innovative ways of reducing lithium battery weight, volume, and cost, as well as reducing charge time to ten minutes for a 200-mile charge.[1] A nickel foil is used to preheat the battery up to 60°C, in 90 seconds, for optimum charge rate. Lithium iron phosphate and graphite replace the costly cobalt. If this can be developed for mass markets, affordable battery-electric cars start to become realistic for the numbers required to help avoid a climate crisis. Cost is king. Avoiding CO_2 emissions in all the processes needed to make the battery and the car it powers is important.

Sustainable liquid fuels

Some applications, such as commercial aviation, using current fleets, require a liquid fuel with high energy density per unit weight, and per unit volume, which is carbon neutral. Some favour a biofuel solution, but this has implications for land used for the growing of food crops. Ideally, we should use CO_2 captured directly from the atmosphere as the basis for such fuels, using renewable electricity to power the processes involved. Technologies for this exist and we will look at these later. Many forms of existing transportation can be adapted to run on such fuels. Many existing cars and vans can run on E85, which comprises 85% ethanol, blended with 15% petrol. Others can be modified to do this. The ethanol can come from atmospheric CO_2, using renewable electricity, but all too often it comes from less desirable sources such as agricultural crops on land that can be better used to grow food.

Solar power is becoming more and more cost-effective. Recent bulk deals for the purchase of solar electricity are already down to about 1 penny ($0.02, €0.01) per kWh. The Hinckley C deal requires 9.25 pence. Offshore wind is around 6 pence per kWh. Solar power from solar farms in Saudi Arabia costs 2.36 cents per kWh based on the deal struck with one of the European utilities. This is equivalent to less than 2 pence per kWh. The United Arab Emirates and Portugal have entered the bulk solar electricity market with prices that are comparable or even a little less, at close to 1 penny per kWh. Even allowing for electrical power losses in the cables and systems needed to get power from such large distant solar farms, solar electricity in the UK is still expected to be the least expensive. It is only at this level of cost that, synthetic production of ethanol from atmospheric CO_2 and water becomes competitive in price.

For the same reasons, the sunbelt is also very appealing as a source for making hydrogen by splitting it from water. It also means that synthetic high-density carbon-neutral liquid fuels should be affordable for the most demanding transport applications. These liquid fuels have the potential to be cost-competitive. It is worth noting that ethanol-fuelled solid oxide fuel cells should provide electricity at 60% thermal efficiency compared with combustion engines at around 35% or lower when measured by the portion that reaches the driving wheels of a car. Moreover, the fuel cell avoids the NO_x problem associated with efficient combustion engines. The emission of harmful particulates will be reduced when fuel cells replace combustion engines. Liquid fuels with a high energy density are suitable for aviation and conventional cars and vans provided the vehicle engines have a multifuel capability. These fuels are of exceptional importance as the world transitions to fully clean power sources. However, they still represent a mid-point on our journey while we build and prepare the means by which most heat and power consumers use green hydrogen, a topic we will continue to revisit throughout this book.

Vehicles powered by fuel cells can be refuelled in a few minutes, just like petrol and diesel cars and vans. Some applications may add a small battery to boost acceleration. Ceres Power of the UK (www.ceres.tech) has exceptional expertise in solid oxide fuel cells and has commercial products on sale now at sizes of real interest. NASA of the United States has solid oxide fuel cells that have a high power density of 2.5 kW/kg, which may be high enough to power small, short-range commercial aircraft with up to 200 seats and a range of up to 2000 nautical miles (NM). This would sit well with the electrification of commercial aviation. In the United States, NASA has been working on fuel cells for many decades to provide power to satellites and sustain life in spaceships.[2] Their work shows progress towards fuel cells that may be sufficiently power-dense to sustain the flight of aero-dynamically efficient aircraft, including some commercial operations in the form of long-range fast aero taxis operating in urban areas with vertical take-off and landing. These could have a range of 500 miles at speeds up to 250 miles/hour. We will look at this later and in more detail in Chapter 10. Promising work on synthetic liquid fuels with a higher energy density is taking place, increasing the chance of increased electrification of some forms of commercial aviation and adapting a wider range of regular cars with a high proportion of liquid fuels. While the electrification of aviation may suit some categories of commercial aviation, there is a need to move to liquid hydrogen-powered commercial aviation, but this will require a radical change in aircraft design which we'll discuss in Chapter 10.

There are some other challenges. Direct burning of a hydrogen fuel may still leave a problem of NO_x. NO_x is formed as a result of time spent at very high temperatures in the burners of a gas turbine. Great progress has been made in the design of low NO_x burners. Fuel cells avoid the exposure of time at high temperature but are still not competitive in terms of being lightweight, thereby reducing the maximum range of commercial aviation if powered by fuel cells.

Many of the older conventional aircraft, such as the Boeing 747, are being retired early, as the demand for long-range travel reduces under the pressure to mitigate climate change, and more modern large twin aircraft are more efficient. Those that continue in service need to use a liquid fuel that has a high energy density to match that of jet fuel but is much cleaner in terms of greenhouse gases.

Clean personal transportation in cars is a realistic goal. It will be paced by the rate conventional cars can be taken out of service or able to use a carbon-neutral fuel. We also need to find ways of manufacturing electric cars using much less CO_2. By 2040, there could be two billion cars on the roads of the world, unless the rate of car purchasing is reduced. If we were to make one billion electric cars in the form we know now, the CO_2 penalty for their build could be around eight billion tonnes. That is equivalent to a 20% increase in the world annual output of CO_2 now! We know how to do much better than that by adding a fuel cell running on carbon-neutral fuel, or hydrogen coupled with a much smaller battery. Smaller batteries need fewer expensive rare metal alloys. Many cars could be smaller, lighter, and more streamlined to reduce the high cost of batteries. There remains work to do to ensure carbon-neutral fuel meets all the requirements. Of course, it needs a new network for the retail of carbon-neutral fuel. In France, some petrol stations have been adapted to sell E85. This is a blend of 85% ethanol, with 15% petrol. Ethanol has problems in that it loves water. The car engine does not share that love. Ethanol is highly flammable. Some people see it as a desirable alcoholic drink! As noted earlier, electric and fuel cell-powered electric cars can be fuelled at home using renewable electrical power, when the value of renewable energy is lowest. Many homes and communities will have their own renewable power sources, complete with a standardized electrolyser and compressor to produce the hydrogen needed to support heat, power, and transport needs.

Direct air capture of CO_2

As noted earlier, even better liquid fuels are being synthesized, with pilot plants being prepared for eventual mass-market roll-out. These synthetic carbon-neutral fuels could be the basis of the fuel needed to progressively replace petrol and diesel fuels in most cars until they are replaced by proper and affordable electric and fuel cell electric hybrid, cars, and vans with the combination of performance and range people insist upon. Without a solution for the 1.3 billion existing cars, the climate benefit from electric cars will be much reduced. Ethanol production using the Oak Ridge ONRL method is being taken seriously by two major oil companies in the United States. Chevron and Occidental are shareholders in a small pilot plant intended to pave the way towards the mass market. Since big oil in the United States has shown interest, ONRL is publishing less about the process in the public domain. For complete success in cleaning the fuelling of residual cars and vans, we need processes like those based on direct air capture of CO_2 and water from the atmosphere as the starting stock for carbon-neutral liquid fuels. Such fuels will go a long way to reducing climate damage from existing forms of transportation, as progress is made to go greener still as befits the urgency needed to prevent a climate disaster.

Direct air capture of CO_2 is one of the most important technologies needed for the manufacture of carbon-neutral liquid fuels. The captured CO_2 is returned to the atmosphere when the fuel is burnt to deliver power. Captured CO_2 has other benefits – it can be processed to make building materials or simply locked away.

Direct air capture of CO_2 must be done in large industrial units that capture more than a million tonnes per year at a price of less than US$100 per tonne. Carbon Engineering Ltd thinks that can be achieved. Their first industrial-scale direct air capture of CO_2 plant will run in 2022. Moreover, they expect to produce a high-energy liquid fuel for transport using the captured CO_2. Others see merit in injecting CO_2 into the cavities below the Earth's surface from which fossil fuels have been extracted. It is hard to see the business case for this route, and that will impede uptake.

Sadly, some businesses see the direct air capture of CO_2 as a useful way of extracting more fossils fuels by injecting the CO_2 into the oil and gas fields to flush out more fossil fuels. Such practices increase the prospect of a climate disaster. However, if it is used to flush out a stream of methane, without leakage, that can be split directly into hydrogen and carbon flakes, the injection of CO_2 might make sense. The hydrogen is what we want for fuel cells. The carbon flakes can be used to make carbon fibres for high-strength structures of light weight. Steelmaking requires both hydrogen for heat and carbon flakes to obtain the right commercial grades of steel.

Carbon Engineering Ltd, located in Squamish, British Columbia, Canada, has published some outline details of the pilot plant that first started to run in December 2017. It is a private company funded by private investors including Bill Gates, Murray Edwards, government agencies, Oxy Low Carbon Ventures, a subsidiary of Occidental, Chevron Technology ventures, BHP, and top-tier Canadian and US government agencies (Figure 5.1).

Large fans draw air over many thin flat plastic surfaces that have a special fluid flowing over them. This non-toxic solution binds with the CO_2 molecules in the air, trapping them in the liquid solution as a carbonate salt. The carbonate solution is put through a series of chemical processes to increase its concentration and purify and compress it, so it can be delivered

Figure 5.1 Carbon Engineering Ltd.'s rendering of the world's largest direct air capture plant.

in gaseous form for long-term storage or used in making a high energy density liquid fuel. This involves separating the salt from the solution to form small pellets. The pellets are heated to release CO_2 in pure form. The small pellets are hydrated in a slaker to be recycled to reproduce the original capture chemical.

Making high-density carbon-neutral liquid fuels for gas turbines in aviation, and renewable petrol and diesel, may require the addition of hydrogen derived by splitting water into hydrogen and oxygen in an electrolyser.

Hydrogen

From my studies, I see mounting evidence for quickening efforts to bring in a hydrogen economy. Realistically, zero-emission personal transportation needs a shift to cars people will want to buy and can afford to do so. While battery-only cars are important, they are not yet affordable by the huge portion of the 1.3 billion car owners of the world. Accordingly, it is unlikely that battery-only cars will make enough difference in averting a climate disaster. Everyone who needs a car and can afford a battery-only car should buy one. Many people making mostly short journeys will find them attractive but for those who regularly have to drive longer distances, they will need something that offers plenty of power, with a long range with fast refuelling – unfortunately, these are not an option with today's battery-only cars. This sets the requirement for hydrogen as the fuel. This will come in time, but in the meantime, there are signs that there are more affordable and practical forms of battery that could result in a bigger market share for battery-electric cars.

Commercial aviation releases most of its greenhouse gases with flights of over 2000 NM. Commercial aviation needs a zero-emission solution for flights over 2000 NM and up to or over 8000 NM. This requires the use of liquid hydrogen. As we'll see in Chapter 10, this, in turn, requires a big change away from tube and wing to a blended wing body (BWB) in order to find the volume needed to carry the fuel without loss of passenger space, plus space for baggage/freight, that is required to make this sector of commercial aviation financially viable. It is likely that very efficient turbofan engines will be adapted to be run on hydrogen. There have been encouraging developments in power with fuel cells with much-increased power density, but we need to go further for fuel cells to displace turbofans with direct burn hydrogen.

For surface transportation, weight and volume restrictions are more easily overcome. Aircraft that are smaller, slower, and designed for short-range flights only are possible applications for fuel cell-powered propulsion. Progress towards this goal has started with the first flight of a small commercial aircraft designed and built by ZeroAvia at their base at Cranfield University.[3] The first flight took place at Cranfield on 24 September 2020. The aircraft was a modified Piper M class fitted with a hydrogen fuel cell and battery pack. Cranfield's Aerospace Solutions assisted by helping them achieve the certification to fly. This is one of the most important milestones in the history of human flight. The impressive teams at Cranfield are playing a world-leading role in the pathway to net-zero-emission commercial aviation working with key industries. A larger aircraft is planned to fly a distance of 340 miles later in 2021 and more ambitious flights by forerunners of feeder airlines will take place shortly afterwards.

Hydrogen is the most appealing form of energy storage, but it too has its challenges. Hydrogen is abundant and there are some obvious applications we can introduce at an early date. With the right level of expert study,

Figure 5.2 Hydrogen fuel cell powerplant used on the first-ever passenger aircraft flight by ZeroAvia at Cranfield University, UK.

hydrogen could play a major role in national economies, while helping reduce the adverse impact of daily life on climate change. It is important for hydrogen to be produced from clean sources. Most hydrogen produced now comes from fossil fuels, thus undermining its attraction as a clean fuel. However, green hydrogen can now be made by the electrolysis of water. It can also be made by bubbling natural gas or methane through a 1-metre depth of molten tin to yield hydrogen and flakes of carbon (Figure 5.3). Both processes use renewable energy.

Some industrial processes require huge amounts of heat, and natural gas is widely used. In earlier times, a large amount of coal was consumed. Some were used directly to make heat and power, while much was used to make town gas for distribution to homes and factories. Many industries could change over to hydrogen gas with a large reduction in greenhouse gas, but it does mean we need to provide an upgrade to distribution networks to carry the gas to the end-user. Hydrogen is the world's best escape artist so steel pipes, as used now, need to be lined with a polymer to minimize the escape of hydrogen. Pipe diameters may need to need to be increased.

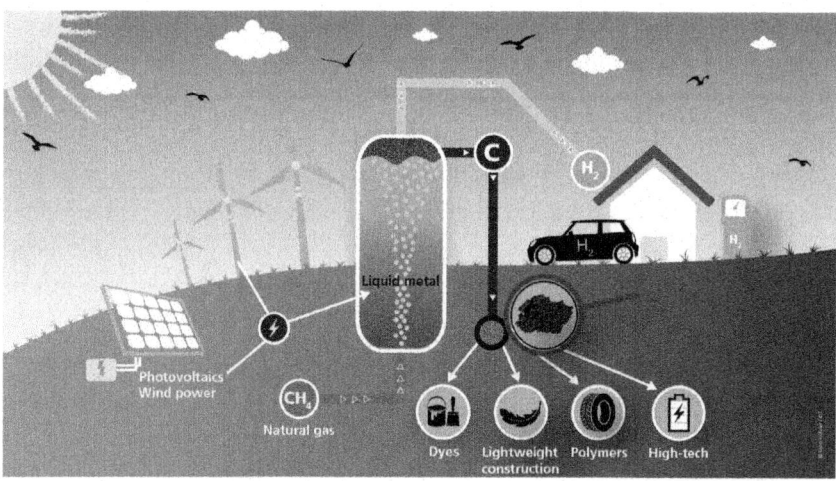

Figure 5.3 Methane pyrolysis by means of a bubble column reactor allows for the climate-friendly use of fossil natural gas.[4]

New-build homes could be heated and powered by hydrogen, derived entirely from renewables. Such homes could collect their own renewable energy, using any excess to electrolyse water to be stored as hydrogen, for use when renewable power is not able to match demand. The combination of renewable power and energy storage in the form of hydrogen, liquid fuels, and batteries will grow in popularity if they can be made affordable. Achieving that requires investment and resources to get to high-volume global markets.

As a fuel, hydrogen has an energy density per unit weight nearly three times that of petrol, diesel, and jet fuel, but for equivalent heat release, it needs four times the volume of diesel or jet fuel, even in its liquid form at the incredibly low temperature of −253°C. While commercial aviation is not the worst offender in the climate change problem, it represents the biggest challenge and provides the opportunity to lead in finding, then applying, the best solutions that will be needed in other markets, notably long-range trucks that must match diesel for range between refuelling stops and load capacity without increasing truck size. The longest-range aircraft equivalents of the Airbus A350 and Boeing 787 at 8000 NM represent the toughest case. Early studies indicate that even the toughest case is attainable. However, it requires big changes to get the hydrogen at the right price together with the infrastructure needed for the fuel supply lines at all the necessary airports. With three-quarters of commercial aviation emissions arising from aircraft flights of more than 2000 NM, it is a priority to find the right solution for long range and we'll discuss this in more detail in Chapter 10.

European energy traders see hydrogen as ten times cheaper to transport than electricity. This is vitally important as renewable energy is harvested far away from where it is used. Hydrogen is set to be the major energy source for industrial and domestic heat and power, and most of our future transport.

Germany appears to be planning to become a global hydrogen leader. Germany, with some of the biggest industries, expects to see investments of about €500 billion by 2050. Siemens Energy sees a strong future in harvesting renewable offshore wind power to make hydrogen by the electrolysis of water offshore then piped to onshore terminals where it will be piped to big industry for steel making and the chemical industries. Some of the renewable power will be transmitted as high voltage electricity. Early plans provide for 100–200 MW of renewable power from offshore wind turbines located in the Baltic Sea and the North Sea; €700 million have been allocated to three companies in Germany to support this work. We can expect to see domestic heat and power growing at the expense of natural gas, coal, and nuclear. In support of this, we need to see the hydrogen gas distribution network grow. There remains the need for detailed work to build the right distribution network and consumer confidence in securing safe and affordable prices. Both objectives appear to be realistic. Many new jobs will be created.

Hydrogen for transport

Hydrogen is abundant and seen as an ideal fuel. It has a high energy per unit mass. It is the only truly clean fuel. It eliminates CO_2, contrails from aircraft, and particulates. For direct burn engines like those needed for long-range commercial aviation, staged combustion will be needed to minimize emissions of NO_x. The tankage for liquid hydrogen at −253°C is more difficult and requires near-perfect insulation for tanks that have a much higher volume than the tanks containing diesel for trucks or jet fuel for aircraft. A means of dealing safely with liquid hydrogen boil-off is required.

Almost 70% of emissions from commercial aviation come from aircraft with a range exceeding 2000 NM. Smaller and slower aircraft, including commercial jets up to about 2000 NM, can use fuel cells. These are very efficient and have no NO_x issues but are not yet light enough for the longest flights. Enormous progress has been made in terms of power/kg and power/litre. Long-range trucks share a need for liquid hydrogen fuelling and can be mutually supportive in finding the means for safe fuelling with liquid hydrogen.

For cars, it is more realistic to use high-pressure hydrogen but the tanks to store the fuel are heavy and bulky. The Toyota Mirai, a hydrogen-powered car, carries 5 kg of hydrogen pressurized to 700 bar (10,000 pounds/square inch). Two tanks are used and together they weigh 87.5 kg.

The newly formed Kubagen company (www.kubagen.co.uk) has developed a new technology that reduces the volume/kg hydrogen stored by a factor of four. This technology allows the volume of the fuel tank to be reduced, making it easier to find the space within the car.

Five kilograms of hydrogen store the energy equivalent of 14 kg of petrol. Better still, if we use a fuel cell to convert the fuel energy into electricity, the efficiency is about 60%, whereas the petrol engine delivers power at up to 35%, but much less goes to the wheels.

When judged by the power at the wheels, the 5 kg of hydrogen could be the equivalent of 40 kg of petrol. This is equivalent to about 53 litres of petrol. There is another attraction for the Kubagen hydrogen storage system in that it uses a lower pressure (less than that of a scuba diver's tank) and has the potential for a five times lower cost per kg stored than the 700 bar tanks. Work continues elsewhere to find the most compact form of storage for hydrogen.

Fuel energy density comparison

Tables 5.1 and 5.2 compare fuels from the point of view of energy density in terms of kWh/unit weight and per unit volume. Hydrogen is great on energy density per unit weight, but it is clearly a non-runner in terms of volume for large conventional commercial jets. However, we can change to a different configuration of aircraft in the shape of a blended wing body that may provide all the volume we need to use liquid hydrogen at a temperature below −253°C. We will return to this later, as it may well prove to be the best way forwards for commercial, long-range flights.

Table 5.1 Energy density in terms of kWh/kg.

Fuel	kWh/kg
Hydrogen	33.33
Petrol	12
Natural gas	14.8
Ethanol	8.3
Liquid ammonia	5.1
Lithium ion	0.1

Table 5.2 Energy density in terms of kWh/litre.

Fuel	kWh/litre
Petrol	9.5
Liquid hydrogen	2.4 (at −253°C)
Hydrogen at 1 bar	0.0028
Hydrogen at 690 bar	1.25
Natural gas	0.01

These figures explain the reason why petrol, diesel, and aviation fuel are so difficult to match with cleaner fuels. Even after allowing for the higher heating value per unit weight, gaseous hydrogen at normal pressure and temperature requires a fuel volume 3400 times greater than petrol for the same heating value. In its liquid form, hydrogen at −253°C requires four times the volume of petrol but note that it takes a lot of energy to liquefy hydrogen at exceedingly low temperatures. There is some relief in that a hydrogen fuel cell might require only half the energy for the same power output, say at the wheels of a car. There is also scope for the optimization of car designs in a way that makes it easier to use cleaner sources of stored energy.

The point is, as we've seen previously, it's pretty clear as to why hydrogen is the clear winner going forwards, but it's going to require a significant amount of design engineering to ensure we can maximize the gain while minimizing the challenges and risks.

Hydrogen storage for transport

Using hydrogen for transport requires moving hydrogen to the point of use. This has been difficult, until now. We can see more energy-dense hydrogen fuel storage, at a lower cost, emerging as a potential solution.

To date, the lack of achieving a high volumetric energy density has held back the use of hydrogen as the clean fuel of choice for transport. So severe are the problems that mobile applications have not reached the market in the volume needed to benefit climate change.

But that may be about to change as a result of real progress in hydrogen storage using transition metal hydride storage that is four times better than a 700-bar tank, with all the problems associated. It also promises to be about five times cheaper than 700-bar tank storage. The new metal hydride sponge system can operate at normal ambient conditions with little or no heat management issues.

Engineers have done their best with the two existing solutions. There are two railway train applications running trials. One uses huge low-pressure tanks. These carry a huge penalty in terms of much-reduced revenue-earning space. The other uses very high pressure tanks. Here, the lost revenue penalty is overcome but the tanks and feeding infrastructure are potentially expensive.

For cars and vans, ready for delivery to retailers, only the high-pressure tank is viable, but we really need something better. Thankfully, it appears that the patented transition-based Kubas Hydrogen Sponge from Kubagen will take hydrogen from the dream clean fuel to reality.

While the work of Kubagen represents a big step in the right direction, and it may be close to the best system nature allows, there is scope for further innovation and work continues with signs of progress.

Imagine the scene! The wind turbines are turning, the sun is shining, especially in the sunbelt, making solar electricity retailing in bulk at below 2 cents/kWh, feeding the electrolysers that split water into hydrogen and oxygen. The hydrogen is delivered to the co-located plant where it is packaged into transportable hydride tanks, ready to be delivered to fuel depots then on to fuel retailers.

Many vehicles will be powered by fuel cell/battery hybrid systems with the fuel cell powering the cruise, while the battery takes care of big transient power demands. Both power sources are clean and the battery size required can be much smaller without loss of performance while achieving a cost reduction.

Using hydrogen in this way creates the prospect for global mass-market electrification of all surface transport. That will help mitigate climate change – as we'll see later, 75% of all transport CO_2 emissions come from road transport.[5]

However, do remember we need several bigger contributions to really get close to what is needed for climate change mitigation. We need manufacturers, fuel retailers, and the public to understand the importance of hydrogen in tackling climate change. We then must optimize the best way for making hydrogen at the right cost using renewable energy, as well as optimizing its distribution and the fuel cell plus battery combinations that meet customer and investor needs for it to achieve mass-market uptake.

Notes

1 Penn State University (2020), *High Energy Li-Ion Battery Is Safer for Electric Vehicles*. Available at https://news.psu.edu/story/610880/2020/03/04/resea rch/high-energy-li-ion-battery-safer-electric-vehicles (Accessed 6 March 2021).

2 NASA (2017), *Space Applications of Hydrogen and Fuel Cells*. Available at https://www.nasa.gov/content/space-applications-of-hydrogen-and-fuel-c ells (Accessed 11 March 2021).

3 ZeroAvia (2020), *ZeroAvia Completes World First Hydrogen-Electric Passenger Plane Flight*. Available at https://www.zeroavia.com/press-release-25-09-2020 (Accessed 8 March 2021).

4 Karlsruhe Institute of Technology (2019), *KIT and Wintershall Dea Launch Collaboration on Climate-Friendly Industrial-Scale Methane Pyrolysis*. Available at http://www.kit.edu/kit/english/pi_2019_141_hydrogen-from-natural-gas -without-co2-emissions.php (Accessed 14 March 2021).

5 Our World in Data (2020), *Cars, Planes, Trains: Where Do CO_2 Emissions from Transport Come from?* Available at https://ourworldindata.org/co2-emis sions-from-transport (Accessed 10 March 2021).

6

RENEWABLES

DOI: 10.4324/9781003193470-8

Petrol rationing was introduced on 13 March 1942 and lasted until 1950. Private use was not allowed until the end of World War 2 in Europe. Dad was then able to take me for rides in the adjacent country.

I saw farmers using windmills to pump water into storage tanks low on the ground to be used for crop irrigation when crops needed it. The embryo design engineer in me noted that when the windmills stopped turning through lack of wind, there was stored water with enough height above the ground for water to flow to irrigate crops in need. There was also water for the animals to drink when the local streams and ponds became dry.

I was about 11 years old when I learnt that renewable wind power needed storage. All renewables depend on storage. Moreover. that storage must be capable of reaching the right place at the right time and in the right form at the right price. At that stage in my life, I had no idea what an engineer was. I wanted to be a woodwork teacher, where I could use my design and woodworking and modelling skills to make things of value, but in the field, the wind-powered water pump was another activity that grabbed my attention. People were flying model aeroplanes that were quite unlike real aeroplanes, and every plane was different. This sowed two new seeds in my mind. People were designing quite technical things that worked. Flying needed many innovations.

All energy on Earth originates from the sun. Our coal, oil, and natural gas came from the radiated energy of the sun, feeding the growth of plants that fell down at the end of their life, and after millions of years, natural processes led to the formation of fossil fuels that we now mine and process into convenient high energy density fuels to provide for our heating, lighting, transport, energy, and so much more.

Renewable energy has some challenging problems. This form of energy is harvested in a very low-density form. Already major utilities are buying solar power in bulk from places in the sunbelt. Often the best places to harvest this energy are remote from where it is needed, and much of it arrives when the demand is low. Together, these mismatches between demand and supply need demand-side controls, and several appropriate forms of storage, in order that the incoming low-density supply can be delivered to the end user in a high-power density form when and where needed. There are many ways of doing this, with some of the many ways having more useful merits than others.

We will look at the strong candidates for energy storage mentioned earlier in more detail later. We will also look at how best to transport and distribute renewable energy from where it is generated to where it is needed.

Most countries need renewable power from wind turbines and solar. Most will have a good reason for nuclear power provided it can get close to matching the cost rates of renewable energy. All three forms of energy generation need energy storage and distribution networks.

Wind

Many countries are blessed with generous flows of wind that can be used by wind turbines to generate renewable electricity for domestic consumption and export, and as with all renewables, it needs energy storage to enable the supply to match demand. The earliest wind turbines, in a more basic form, have been used purposely since the late 19th century. Now many countries have a significant dependency upon wind turbines putting large amounts of electrical power into national distribution networks, such as The National Grid of the UK.

The UK is the windiest country in Europe, with 40% of Europe's wind energy blowing over it. The bulk of UK wind power comes from offshore installations, with the majority being planned to be located off the coast of Scotland. These turbines are much larger than those onshore.

Offshore wind costs £120/MWh in 2018, but it is forecast to reduce to £58/MWh for installations scheduled for completion in 2022. There remain concerns about maintenance costs in the very hostile offshore wind farm locations. There is the long-term experience of offshore oil and gas rigs to guide offshore wind turbine designers and operators to help them in controlling maintenance costs. Onshore turbines cost £65/MWh, with a reduction to £46/MWh for installations scheduled for completion in 2022. In 2019, 14.8% of all UK electricity came from wind turbines. Plans exist to increase this rapidly, but upfront costs do moderate the pace of completing new installations. In November 2019, the average UK power demand was about 41 GW with renewable wind contributing 10%, and solar contributing 3%. While wind turbines are suitable for grid supply, the picture becomes more complicated when the necessary energy storage costs are added in together with the high costs of new power transmission lines. However, the need to decarbonize everything using renewable energy provides a strong potential market for batteries, hydrogen, and carbon-neutral liquid fuels with a high energy density.

Each of these storage systems will require huge investments, but most make a strong business case that fully justifies the investment. Much work has already been done to optimize the best processes. Judgements are being made about the pace of uptake in reaching all applicable markets.

We already know that the world needs huge amounts of all three forms of renewable energy storage.

Using the UK as a worked example, excluding wind and solar, the UK Electricity Grid contributed 295 g of CO_2 per kWh. With 2019 values for wind and solar this reduces to 288 g of CO_2 per kWh. The UK National Grid publishes its output and CO_2 performance for all to see.[1] Much detail is provided enabling anyone to see how much comes from each source. The average CO_2 figure for 2018 was 270 g/kWh.

Hornsea Project Two, Wind Turbine, due for completion in 2022 will be the largest in the world with a capacity of 1386 MW, enough to power 1.3 million of our 27 million UK homes. The UK's largest onshore wind farm is Whitelee in Scotland with 215 Siemens wind turbines providing a capacity of 539 MW.

Many people are opposed to onshore wind and solar farms, mostly because of their visual impact on the scenery that surrounds them. However, we need to try to achieve the right balance between our requirement for our beautiful natural landscape, and our standard of living, as well as avoiding a climate disaster. There are strong plans to plant large numbers of trees to increase the take-up of CO_2, and these can be used to enhance natural scenery, as well as restoring habitat for nature's insects and animal life.

It might be better to have more land-based wind turbines generating electricity to make hydrogen go into national natural gas grids, such that the heating of our homes can create less greenhouse gas. Industry also uses a lot of gas. Having a higher hydrogen content per unit heating value would reduce their greenhouse gas contribution to climate change. It may be more affordable to produce hydrogen on land rather than at sea when using wind turbines as the renewable electrical power source. However, the sums need to be done to prove that. Making green hydrogen offshore is possible, in fact, Siemens Gamesa and Siemens Energy have announced a project to do just that which they hope to have a demonstration operational in 2025/2026.[2]

The increasing urgency to avoid a global climate disaster provides real pressure to move to an economy based on hydrogen (a point we will come to in significant detail later). We need to gain the understanding of everyone that this is necessary and desirable. This will take real effort, especially in industrialized nations, as few of the public understand the threat of a climate disaster until it hits them as individuals. We need to listen to public

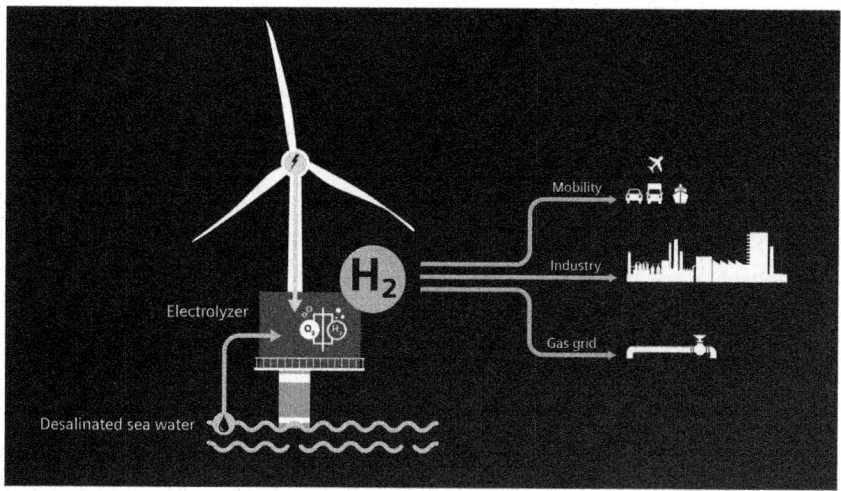

Figure 6.1 Siemens Energy and Siemens Gamesa plan to build a new era of offshore
      green hydrogen production.

concerns and answer in a way that people will trust. There are many factors
needing proper solutions, with a rationale that is easily understood.

The need for energy storage for wind turbines is easy to justify. During
a typical year, the average output of wind farms is 28% of installed capacity for onshore and 39% for offshore. Onshore there are 1516 wind turbine projects with 7047 wind turbines connected to the UK National Grid.
There are 32 offshore wind farms with 1716 turbines connected to the
National Grid. With these low outputs as a percentage of installed capacity,
energy storage is vital.

With the strong winds for a 24-hour period in November 2019, the
wind turbine average input to the grid was 17.2%.

Almost all UK solar and a significant amount of wind turbine power is
connected to a transmission network, rather than the National Transmission
Grid. Many other countries will have a similar set up of local area networks
sitting alongside a national electricity grid.

For reference, during the 7–8 November 2019 period, nuclear power
delivered 16% of demand, while coal still provided 1.6%. Mainstream supplies are dominated by combined cycle gas turbines. Perhaps some of these
power stations might, in the future, use a carbon-neutral fuel. This might
be realistic, in transition, as the UK and other similar nations strive towards
the goal of net-zero carbon.

For most countries, the relatively low average output as a percentage of installed capacity for wind turbines and solar panels is a serious issue, as there will be days when output from either or both will be even lower. There must be sufficient energy storage to enable the demand to be met. Turning off the least critical electrical loads can help in keeping all essentials with a dependable continuous supply. All these good installations place the UK as a front runner in renewable energy from wind turbines. Germany and Denmark are also making good progress, but all these excellent installations fall well short of what the Earth requires to alleviate adverse climate change damage already done and being added to at a fast rate.

Solar

Unlike those near the sunbelt, many countries are less well placed to have a high dependency on solar radiation harvested on national land and sea resources. The radiation must penetrate a thicker atmosphere due to the shallower angle of incidence from the sun. More frequent and thicker layers of cloud erode the yield still further. Despite this, solar power is worth the investment, and it does help keep the lights on. Much of central and northwestern Europe is within reach of the Sahara Desert using high voltage DC power lines or by converting atmospheric CO_2 and water into a liquid fuel that can be transported, as oil is now, for use when and where needed. Even more important is that the cheapest electrical power will be used to make gaseous hydrogen that can be piped into a hydrogen gas grid the serves the whole of Europe. There are plans for a subsea pipeline from north Africa to Italy, then onwards to the whole of Europe. Moving energy is ten times cheaper by hydrogen than by electricity.

There are other hot deserts, although the Sahara is the largest of the hot ones. There are other countries closer to the UK that have a higher potential for solar power. Spain and Portugal are examples. Liquid fuels derived from atmospheric CO_2 and water, powered by renewable electricity, can be used to power the supertanker bringing such a fuel to the nations needing that fuel to replace petrol and diesel fuel used now. The United States is well endowed with large areas suitable for large solar power generating farms, many of which are used to extract oil and gas. The demand for these fossil fuels must be curtailed by massive amounts, as the United States is one of the big contributors of greenhouse gases, accelerating the slide towards a climate disaster. The United States also has a high potential to be one of the powerful leaders with actions to help avoid a climate disaster.

As we've seen, renewable sources of energy have been with us for a while and yet they still haven't been able to scale to deliver enough impact on reducing our emissions. But this might be about to change.

Can renewables compete with nuclear power?

One of the frequent criticisms of current renewable energy generation methods is that they cannot match the output of current methods of electrical generation such as nuclear power.

Given recent developments and some novel approaches, I think this is an issue that will cease to be true. Allow me to demonstrate by investigating if it's possible to build a solar power station that could match the output of a modern industrial-scale nuclear plant like Hinckley Point C.

Saudi Arabia already recognizes that it has huge potential for solar-generated exports of electrical power, but there are other countries that are also well blessed with solar radiation, and some in great need of solar power for both domestic use and as a source of revenue from exporting renewable electric power. This is especially important for the countries supplying fossil fuel in huge amounts to meet high demands for petrol, diesel, and aviation fuels. The United States is an example.

We need to look at costs for producing renewable power. We are already seeing the bulk purchase of solar electricity, at prices below all competition, at levels of little more than 1 penny per kWh. EDF has struck a deal in Abu Dhabi for a 1.5 GW supply at a price of $0.0135/kWh. Of course, there are costs to be added to achieve a dependable supply at the point and time of demand. These additional costs are more than recovered by the sales of stored energy, as noted earlier.

Solar farms have a large footprint, but there are enough suitable sites to meet demand. Solar farms in the poorer countries receiving high solar radiation could transform a life of poverty into one very much better. As an example, we can consider building a new solar power station in Mali, the eighth-largest country in Africa. It is a land-locked country of 1,240,000 square kilometres, with a population of 19.3 million. It is noted for its crises of poverty, starvation, and disease with little outside help. This country sits in the sunbelt, receiving up to 1 kWh per square metre, at ground level, for up to 12 hours per day. Much of this country is the Sahara Desert, meaning that most of the land is hostile and presently without good use. To illustrate the scale of a solar plant in Mali to match the output of Hinkley C, we can calculate the size and value of the solar plant and its output. Hinkley C will have an output of 3200 MW. In one year, its output from running continuously at rated power is $3200 \times 365 \times 24 = 28.032$ million MWh. For the solar panels receiving 1 kW/metre squared, the output with current solar panel technology is 220 watts/metre squared. Assuming we achieve ten hours per day then the annual output of electricity per square metre is 803 kWh. If we take all this solar power as electricity, the solar panel area must be 28,000 million kWh, divided by 803 kWh per square metre which equals 35 km². This is a square of about 6 km side length. Realistically, the output per square metre might be only half due to restrictions on the orientation of the solar panels. This would result in the need for a panel area of 10 km square. Noting that we need energy storage, and we need to replace fossil fuels with fuels that are cleaner, in terms of greenhouse gases, we might find it advantageous to use some of the intermittent renewable electricity to make cleaner fuels such as hydrogen or carbon neutral liquid fuels.

For example, we might retain 60% as electricity and use 40% of this electricity to power a process to make ethanol, a clean liquid fuel. The efficiency of the conversion is about 30% so the net output drops to 578 kWh. The area required to match Hinckley C rises to 48.5 km squared. This requires a square of a side length of 7 km. Hinckley C will provide about 7% of UK electrical power so we would need about 14 such sites of 6 × 6 km, or 10 km² if we use the more cautious figure. This could easily be found in the solar sunbelt. With a transition to more power derived from solar renewables for surface transport and for aviation, we might need to double this amount. This should be no problem, beyond that of doing the right deals with landowners in the sunbelt. It will need some large and long high voltage transmission lines.

The energy traders of Europe have plans for a supply of renewable energy covering the whole of Europe, including the UK. Much of this originates as solar power farmed in the sparsely populated areas of North Africa, where solar radiation is relatively intense, and there is a need for locally generated revenue to uplift living standards locally.

The energy traders calculate that it is cheaper to transmit energy as hydrogen, rather than electricity. They calculate that transmission using hydrogen is ten times cheaper than transmission by electricity. This fits well with a complete changeover to heating and power using hydrogen, and powered transport using fuel cells, coupled with batteries that are well suited to the role of transportation. There are many applications for hydrogen fuel cells that can be put in place before countries have a hydrogen gas grid. Fuel cells can be run in reverse to make hydrogen using electricity when it is at its cheapest.

Some countries already have this differentiation between night and day; encouraging consumers to shift some consumption to night-time periods when demand is less which helps the suppliers avoid overload during peak times. Countries moving to increased use of renewables might choose to employ tariff moderation to suit the variability of supply from renewables. Countries can use night rate electricity to replenish the hydrogen fuel tank of a car, bus, or truck powered by a fuel cell. This would suit someone with off-street parking and daily mileage that comfortably fits the energy storage limits of the energy storage system of the car. There are plans by the petrol/diesel fuel retailers to equip well-chosen petrol stations with hydrogen dispensers.

As it was in the early days following the introduction of unleaded petrol, long journeys will involve some planning until more complete coverage becomes economic. However, for many users, with off-road parking, the recharging of energy stored in the vehicle can be done at home or the base depot.

Possible problems with solar farms in hot deserts

Is it going to rain in the Sahara?

There are potential problems with huge solar farms in the Sahara Desert that need to be understood and safely controlled. Huge areas of solar panels alter the solar energy absorption of the desert. More solar radiation will be absorbed by the panels, such that there will be some cool areas that were once hot. This may cause mass vertical movements of air. The air above the desert contains moisture averaging about 25%. When cooler air is pushed upwards, it could result in rain falling, thereby changing the nature of the desert floor. While this might be most welcome, it could be a big change for the desert climate. It needs careful thought with accurate modelling, plus the study of relevant natural arising. Understanding how tropical rainforests work might help answer this question. So far it appears that solar farms in the sunbelt have escaped these issues. In engineering, unlike public life, everything is guilty until proven innocent. Do we have the proof? The change is likely to be a big factor in designing homes, industrial plants, and infrastructure needed to support the solar farms. Huge solar farms have been built without seeing signs of this problem. As more large solar farms are added, we should look for early signs of problems.

It is well known that desert winds cause drifting sands to form dunes that move over time. We know from our experience of drifting snow in the Derbyshire hills that tall drifts form near dry stone walls. Drifts of snow five metres high can form quickly near the walls of stone that are barely a metre tall. If drifting sand dunes were to cover the solar panels, there would be a loss of output and there would be a risk of damage.

All possible hazards must be identified and provided with tested solutions. All renewables present stiff challenges, and all these need proven solutions. Mother Nature has taught us many lessons relevant to what we, as humans, are trying to do for ourselves to help achieve sustainable life on Earth. We must heed these lessons and use them constructively, very quickly. More than this, solutions are needed at a mass-market level, not just in laboratories and small-scale pilot plants.

Summary

As industrialized countries, such as the UK, increase their use of renewables, they will import renewable energy from the lowest-cost sources overseas, such as the Sahara Desert, where population density is low. Middle East countries, like Saudi Arabia, supply much of our fossil fuels. Saudi Arabia is richly blessed with solar energy. They well know that as sales of fossil fuels decline, renewable energy sales can replace that income. Europe could take it in the form of electricity, or more cheaply, as gaseous hydrogen. It could be processed into a high energy density liquid fuel to be shipped as oil is now, in tankers cleaned out and refitted to take the new fuel. As we've seen, European energy traders have thought through plans for a European hydrogen gas grid.

Importing solar energy from the sunbelt is realistic. The Sahara Desert covers 9,200,000 square kilometres. That is 37.5 times the total area of the whole of the UK. It is equivalent in area to the whole of Europe. After energy losses through the atmosphere, each square metre of surface receives about 1 kW/hour of solar energy. With the current technology, solar panels can deliver 220 W to the grid per hour for every square metre. The present world requirement for electrical energy could be met using present-day solar panel technology with panels covering just 2% of the Sahara Desert.

Sounds easy, but the engineering is tough, especially when you need to add in the essential energy storage for a total cost that consumers can afford. As we've discussed before, the replacement of all fossil fuels with renewables significantly increases the world's demand for electricity. Increased energy usage might add significant demand, but as I've demonstrated, it is still within the reach of renewable energy sources from the hot deserts of the world.

Notes

1 National Grid ESO (2021), *Welcome to the National Grid ESO Data Portal.* Available at https://data.nationalgrideso.com/ (Accessed 11 March 2021).
2 Siemens Energy (2021), *Siemens Gamesa and Siemens Energy to Unlock a New Era of Offshore Green Hydrogen Production.* Available at https://press.siemens-energy.com/global/en/pressrelease/siemens-gamesa-and-siemens-energy-unlock-new-era-offshore-green-hydrogen-production (Accessed 14 March 2021).

7

(DECENTRALIZED) NUCLEAR POWER

DOI: 10.4324/9781003193470-9

My main career background is in designing aeroengines for wide-body jet airliners. Here, high performance is a must. In turn, the designer must extract the greatest strength material can deliver through a wide range of operating conditions. Every part must work close to its maximum performance capability. Even a small drop in performance signals the need for corrective action. As noted earlier, we also know that there are life-threatening failure modes that can happen, so we must make sure that this never happens, but if it can happen, then it will happen. When this happens there must be a safe, and well-proven solution, that makes sure that no life is lost.

The above scenario is a typical situation in aviation design engineering. Imagine an engine that has been on the wing for five years and is starting to lose a tiny amount of performance. Many instruments on the engine will pick this up but more than that, these instruments will know exactly which component is starting to degrade. The engine is modular so the flight operation controller will find the right location and time for the aircraft to come out of service for a few hours while the module with the suspect part is removed and replaced with a healthy one. The changeover usually takes place at night. Because of the fact that all modules are physically and functionally interchangeable, there is no need to run the engine to check. We know it will work properly, so passengers are loaded and off they go to their destination in complete safety. The revenue is retained. The cost is minimized, and safety is assured. No one has been alarmed or concerned because everything has been safely controlled. This is how it must be as huge numbers of people reset the way they live, work, and play.

Wind turbines and solar power are valuable sources of electrical power, but they can never be relied upon to keep everyone with dependable power 24/7, so in addition to adequate energy storage, we may need another major source in the mix. Presently, combined cycle gas turbines and nuclear fission are the best proven solutions. Conventional wisdom pushes nuclear power stations to be ever larger. There is evidence to show that the way nuclear power stations are designed and built may be improved upon. Further research led by Tony Roulstone at Cambridge University suggests that smaller factory-made units can be cheaper. Moreover, the more you make of a specific design, and built using the same proven materials and processes, the cost per unit output of electrical power reduces. Achieving low-cost electricity from nuclear power requires two absolute essentials. Factory-made small standardized modular units require a launch customer platform of hundreds of units. Moreover, the build time must be fast in order to reap the benefit. Big bespoke units built on site cannot meet these requirements. The investment is large, and the payback time is far too long. Worse still, the big units carry more risks in regard to failure and radiation leakage.

In the UK, it is hard to see a case for Hinckley C with its guaranteed unit cost of about nine times that of the cheapest solar power from the sunbelt. It may be hard for any industrialized country to compete in the global marketplace with the high cost of electricity guaranteed from large bespoke nuclear power generators. Using many more factory-built small modular nuclear generators needs a multinational programme. For small factory-built nuclear generators of electric power to reach the right level of price/kWh to the consumer, there must be sufficient launch customer to justify the high initial cost to manufacture and install more than 100 units each with an output of 300–500 MW. Production rates must be fast to repay the high costs of investment. This should be possible, as many nations will find it hard to reach net-zero emissions from their own domestic renewables. Affordable nuclear power, as part of the power generation mix, might suit many nations. Large bespoke nuclear reactors that take many years to build will always result in electricity selling prices that erode national competitiveness for products and services in the global market. Nuclear power is important, but the cost must be right. There are issues of security, leading to a need for domestic energy-dense forms of power generation. This might best be provided by many small factory-built nuclear power generators, to be located close to centres of high demand for electric power and stored energy for domestic, industrial, and transport applications.

We need a new breed of smaller reactors with fail-safe modes that fully meet all the worst possible forms of adversity. We can now properly model all such events and design them to achieve safe failure modes. The airline industry uses this way of avoiding life-threatening potential failures. Many small modular nuclear power stations might add robustness to power security.

In the interest of national security of energy supply, nuclear power might well earn an important role in the mix of a nation's power generation. Many countries might find the factory-made standardized small modular nuclear power generators well suited to their domestic energy security. Moreover, units might be tailored to match the demands of container ships and cruise liners. For nuclear power to be part of any national power mix, the "fear factor" many citizens have about the safety of nuclear power must be addressed. It helps to know that UK-designed nuclear power stations have proved to be safe, unlike some other designs.

As an example of progress with small modular reactors (SMRs), we should look at the NuScale Power solution, in the United States (www

.nuscalepower.com). Their plan appears to fit the potential requirements for clean, safe nuclear power generation in many countries.

NuScale Power solution

With the first-ever SMR to receive US Nuclear Regulatory Commission (NRC) design approval,[1] NuScale is bringing in the first SMR power plant online in the United States this decade. Their SMR technology, the NuScale Power Module, can generate 25% more power at 77 MW of electricity resulting in a total of 924 MWe for their flagship 12-module NuScale power plant. They also offer smaller units with four or six modules. Their technology permits shorter nuclear construction periods at under 36 months, with lower costs.

The innovative design by NuScale has created a new standard for rigorously proven safety through the Triple Crown for Nuclear Plant Safety. The plant shuts itself down and safe cools indefinitely, with no operator action, no AC or DC power, and no added water. A first for commercial nuclear power.

Other companies are working along similar lines. The overall message is clear and SMRs deserve to be part of national plans for clean electricity. We do need to clean up the emissions released in making the steel and concrete used but there are already established ways of doing that.

Small nuclear power generators built in factories

We must find out what can be done with small modular nuclear reactors to match the output of one pair of reactors, such as those that comprise Hinckley C. This output could be generated by ten such reactors, each running at 300 MW output. Proper work by experts taking a total system approach is needed. Those skills are available but need to be pulled together. The reduced amount of flying may free up design, manufacturing, and installation capacity for aircraft and aeroengine making companies to make the large numbers of SMRs required. The skill base is appropriate, but it needs serious funding, at a time when national debt in many countries is high and rising sharply. As with NuScale, we might want the basic reactor smaller still such that we could think in terms of fully automated mass production with thousands made each year. Modules of multiple reactors could match specific needs. Production in high volume is the route to the best performance and price.

Rolls-Royce builds smaller nuclear fission reactors to power a small number of submarines, using expertise derived from Westinghouse of the United States, and they are using this experience to take a leading role in the development of small nuclear power generators. Numbers are small so the price benefit is reduced. If these units were adapted for commercial use throughout the UK, the numbers could be in the hundreds. Exports could further increase the number. It is essential that modern design processes are used with comprehensive and accurate modelling to show complete safety for every worst possible threat. Installations provide safe solutions for the worst extremes of natural disasters, and drastic acts of terrorism. These extreme design considerations erode some of the cost benefit, but in troubled times the additional resilience against adversity is justified. We do this now for every commercial aircraft and its systems and engines.

However, hard-won global manufacturing experience requires big volume throughput rates to get the best performance and price.

These small nuclear reactors could be used as the source of electricity for local consumers, and as the power source to make renewable fuels for transport and other applications. This much bigger number of units, coupled with better design, build, installation, and operating procedures would lead to lower costs per kWh.

More than this, there is a double safety benefit. Each variety would have its own type certificate. This says all units are the same, which means that precise testing only needs to be performed on a small sample to prove the absence of any defects that could cause harm before its next scheduled check. There are good reasons why each unit will have a safer failure mode, in that each one carries an adequate amount of water to prevent a runaway event leading to the discharge of harmful radioactive material, even if all power to the nuclear station is lost. With numbers of reactors in the hundreds, or even thousands, it becomes realistic to use safety management techniques used by the airline industry. Every aircraft and its engines are built in a tightly controlled way that is fully documented and kept in an easily accessible way. This means that at the first indication of an emerging problem, engineers can quickly identify other aircraft or aircraft engines that may be at risk. Immediate action can be taken to ensure that the severity of the operation of a suspect product is reduced, restoring safe operation, albeit with a shorter life and higher cost, until a thoroughly cleared modification allows restoration of full output without loss of safety. This is a well-proven practice in commercial airlines all over the world.

It is almost beyond belief that one of the most trusted companies, Boeing, has slipped from grace, harming the sound reputation of properly trained engineers and operating crews. As one of my design engineering bosses repeated on and on, "Beware the accountants". The builders of large bespoke nuclear power stations are certainly as thorough as best practice in aviation, but every unit is unique. In the event of a cause for concern, action must be taken to avoid the concern becoming dangerous, and this may require a significant loss of power for a long period while the problem is understood, and a dependable fix installed to allow full power to be restored. By having factory-built, mass-produced nuclear reactors, we can keep the safest running at full output while reducing the power of a few that may be affected by the problem. This retains most of the revenue and gives time to ensure the cure is both effective and safe to return to a full load.

The UK's nuclear power track record is top class, but global experience leads to many distrusting the safety of nuclear power. I know my view may be out of line with the views of experts in nuclear power station design, but I feel we need a step up in the perception of nuclear power station safety, and we must lower the cost per kWh. I would strongly advocate a multinational approach as many countries face the same issues of achieving safe clean affordable electric power 24/7 with a long service life of many decades. The issues holding back investment are dominated by the fear factor, plus the high upfront cost of getting the first few units into service. The United States and China are actively pursuing advanced nuclear power station designs. We can build safe and more affordable nuclear fission power stations, and as a clean low carbon source of electrical power, their high-power density makes them attractive as a safe and affordable source of electric power, with many advantages over farmed renewables with all the added complexity and cost of energy storage and long transmission power lines. Ongoing political tensions point to the need to include nuclear fission power generation in our mix of suppliers meeting a growing demand for electrical power.

Nuclear fusion

Scientists have been able to understand how the sun makes vast quantities of energy leading to a desire to engineer a system using the same understanding that will give us limitless power without the nasty side effects of present nuclear fission power stations. The sun is mainly composed of hydrogen gas but in a most unusual form. The hydrogen is compressed to extremely high pressures under the forces of the sun's own internal gravitational attraction. This gives rise to very high pressures and temperatures in the sun's core. The sun has a radius of 432,326 miles, while Earth has a radius of about 4000 miles. The sun is bigger in diameter by a factor of over 100 times. This means that the gravitational pull is more than one million times greater, giving rise to the fusion of hydrogen atoms to form helium combined with a massive release of energy that radiates outwards from the core to the surface of the sun and beyond. As it radiates outwards beyond the surface of the sun, the energy reduces in density but still gives the surface of the Earth in the sunbelt at a rate of about one KW/hour per metre squared, in daylight hours. Fossil fuels were created using energy from the sun. These fuels were created from the remnants of things that lived thousands of years ago. Strong efforts to recreate mechanisms used by the sun are starting to show promise with a massive pan-European activity called ITER (www.iter.org).

Lockheed Martin was awarded a contract to produce a compact nuclear fusion power generator with an output of 100 MW that would fit on a truck. A spin-out from the UK Atomic Energy Authority, called Tokamak Energy (www.tokamakenergy.co.uk), also has a programme to achieve a similar result. Tokamak Energy has shown its design works, but so far, the input of electrical power exceeds that of the output. Expectations remain high that small nuclear fusion reactors will give limitless electrical power without nasty emissions or by-products. The challenge ahead remains severe.

A clean grid needs nuclear power

From all my studies and taking a design engineers view, a clean grid might feature a significant dependency on nuclear power as there appears to be a pathway to mass-produced small modular reactor power stations that can sit well with renewable wind and solar where there is a national shortage of home harvested renewables, delivering a unit price for electricity that competes well with renewable prices. As we eliminate fossil fuels, the demand for secure clean energy rises substantially. Just to give some idea of mass-produced "micro-modules", the UK might need a few hundred. A multinational programme might break into the thousands. With volumes like that, a significant level of automation would earn its keep and the price would take a big tumble.

This is perhaps a new concept or at least an extension beyond current thinking, but it is worthy of study by intermediate volume production experts. With aeroengine production recovery sometime away, there is a design and manufacturing base capable of addressing most (but not all) of the skills needed.

There are also some weighty issues that need political debate and a clear understanding of all the factors that need to be balanced. Nations must decide the balance between security and economic competitiveness in the global market. Based on present figures, solar-powered electricity in the sunbelt appears to be the cheapest, but it needs energy storage on a massive scale. Happily, the energy storage systems have the potential to be the basis of many sound businesses. Moreover, the cheapest involves long supply lines, with a high dependency on good international relationship over many decades. History shows that such diplomacy is not always the case, especially around such difficult and sensitive topics as nuclear. Security, coupled with the need for nations to create an income to meet the rising expectations of all their citizens, may give rise to including more nuclear power, with smaller power stations placed near big population centres. There are also issues concerning where energy storage equipment is located. Micro nuclear modules might meet much of the modulation needed for generation to match demand. Removing fossil fuels requires energy storage in vast amounts and where much of it must be transported. The new clean power supply lines must fulfil this demand.

People must feel safe about nuclear power. It needs to feel at least as safe as travel by air is now. Safety requires that we design, build, install, and operate each and every nuclear power station to fully certificated methods. Modular build in factories able to produce in quantity should be a better way forward than the bespoke processes used now.

Provided factory-built small nuclear reactors can produce electricity, at prices per kWh that compare well with onshore wind, a clean national electricity grid may need many small nuclear fission reactors to reliably deliver up to about 50% of average demand. Nuclear reactors can be modulated to reduce power when supply exceeds demand in a predicted way. This might be used to reduce the storage needed to support renewable power. Each country would need to model the ongoing pattern of supply and demand to get the right mix of electrical power supplies to each national grid. Some will choose to divert some excess power from nuclear into transportable stored energy. Renewables need large amounts of energy storage, some with a rapid start-up rate, and others that store large amounts of energy for periods of weeks, before being called on to deliver what the renewables are unable to supply directly. My figure of 50% needs optimizing. If any nation seeks a high level of self-reliance, they may choose to have a high dependency on nuclear. If a nation sees strong export potential for mass-produced small nuclear that would also add weight for a high dependency on nuclear. However, if small nuclear cannot be competitive on price with the best renewables, the balance favours a lower dependency. Each country will need to do its own sums to guide the choice of power suppliers. Once safety is secured, cost is king, so for nuclear to allow countries to have competitive costs for energy, nuclear must compete with onshore wind, and subject to more detailed analysis, that should be possible, remembering the conditions that must be met.

The balanced mix of energy sources must be compatible with the demands of climate change. For example, the UK needs a big reduction in greenhouse gases from its national grid despite having to generate increased amounts. Eliminating CO_2 from the UK national grid represents about 10% of the total annual CO_2 output from all sources in the UK. As we eliminate the use of fossil fuels and increase the electrification of cars, buses, trains, and more, the demand for electricity from the national grid will increase by large amounts, leading to an increased need for a secure clean grid. The present 10% could more than double unless we increase the content from renewables and nuclear with the necessary energy storage.

Climate change needs governments to work with a wide group of experts in a way that carries the citizens and international investors with their decisions. Nowhere will this be as important as exploring the role and opportunity that nuclear power may play. Thankfully, this is a conversation that is well underway but more needs to be done by designers, scientists, and politicians alike to ensure the opportunities and risks are understood by all.

Note

1 NuScale (2020), *NuScale Power Makes History as the First Ever Small Modular Reactor to Receive U.S. Nuclear Regulatory Commission Design Approval.* Available at https://newsroom.nuscalepower.com/press-releases/news-d etails/2020/NuScale-Power-Makes-History-as-the-First-Ever-Small-Modular -Reactor-to-Receive-U.S.-Nuclear-Regulatory-Commission-Design-Approval /default.aspx (Accessed 11 March 2021).

8

BUILDING A CLEAN GRID

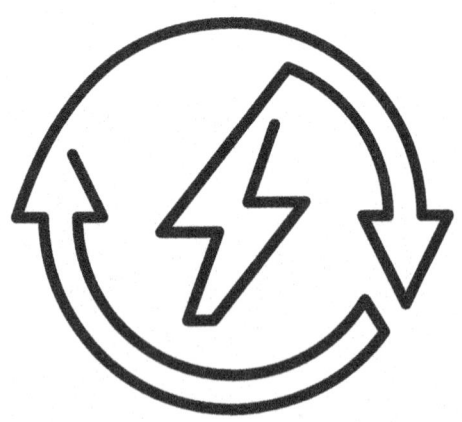

DOI: 10.4324/9781003193470-10

While taking an evening walk along the beachfront at Hove in West Sussex, I calculated the power of the waves at the water's edge and realized how many miles of beachfront it takes to boil a kettle for a cup of tea. There are bigger waves, and we have wind and sun, but the message was clear, harvesting renewable energy is like farming on poor land. More challenging still, renewables take long breaks, so without back-up, the lights go out in the operating theatre! The message was clear, renewables must have sufficient storage to enable all demands to be met. Moreover, energy storage from harvested renewable energy must reach the end-user in the right form, at the right time, and at the right price. Can we secure enough of the Earth's surface to harvest all our needs and how best do we store energy? Sixty-five years later, as I write this late on a windy afternoon, GridWatch shows that renewables are delivering 44% of total UK demand. Another 33% comes from the controllable burning of fossil fuels in combined cycle gas turbines. When these are taken out of service, energy storage will be required. As we remove fossil fuels for all uses, renewables and energy storage must increase substantially. With this dependency, we must get energy storage right. Note that mass-produced micro-nuclear, as yet unproven, might do a very good job in producing stored energy at the right price while creating many good high-value jobs despite all the automation needed.

For many countries, renewable energy needs energy storage that is transportable and as we've seen above, hydrogen is a strong contender for the transport of much of the clean stored energy, but in reality, we will need all three forms, i.e. batteries, clean sustainable high energy liquid fuels, and hydrogen.

While we must start now with the thorough preparation for the ultimate solution of hydrogen, we also need to work in parallel to develop and deliver high energy density liquid fuels, derived from renewables, to be available in global markets at the right price. Doing this will help buy us the time we need to develop and deliver the ultimate solution, failing to provide such a stop-gap simply means by the time we are ready, the damage to our climate will be irreparable.

Citizens and businesses of the world collectively own huge quantities of assets, with a long productive service life ahead. It is not realistic to destroy them all and instantly replace them with new, cleaner assets due to the cost and emissions. Neither is it the right plan. We need to extend the life of the assets we have, using the cleanest fuel possible as far as it is safe and economic to do so. Ideally, products should be designed to last, and fitted with the means for easy upgrades for cleaner use in service wherever that is realistic.

We need to consider the CO_2 penalty for making our products such as homes, vehicles, food, batteries, infrastructure, digital technology equipment, and so much more. This requires the design of things to last longer, with features that make upgrades practical and cost-effective. Single-use fashion illustrates the worst possible practice. Many companies use business models featuring a short product life to generate replacement sales.

Building a clean grid

National electricity distribution grids have a tough challenge in balancing supply to match demand. It does this job well, but the task is much increased as the contribution from renewables increases, and the demand for increased electrification, plus synthetic carbon-neutral liquid fuels and hydrogen. Over time, national grids have developed many tools to keep citizens in comfort and is developing further controls to meet the challenges associated with increased demand, demand variations, and massive variations in renewable power generation.

In my home region of the UK, we do get power outages. This emphasizes the need for increasingly powerful tools to keep the essential services running safely at full demand. The UK's National Gas Grid is demonstrating steel distribution pipes lined with a polymer that might allow hydrogen to be distributed from source to consumers without costly and potentially dangerous escapes of hydrogen.

Tariffs and smart metering are used to reduce peak demand. Back-up energy storage devices such as batteries, carbon-neutral liquid fuels, hydrogen, and hot water storage are used to smooth the supply in meeting demand. As the number of electric cars driven by batteries and fuel cells increases, we can expect to see these car batteries, and fuel cells, used as part of the energy storage system.

Flexibility services, in which homes and businesses become suppliers to the grid, form part of the security plan for any national grid. Their assets for electric power generation are brought in to top up the grid when needed. The national grid pays for the measured input. In the UK, it is predicted that by 2030, 20% of the UK energy system will be provided by these flexibility services. This tool becomes more important as the grid increases supply from renewables.

Demand turn-up (DTU) is used to encourage large energy users either to increase demand or reduce generation when high renewable generation exceeds demand. This happens overnight and in afternoons in the summer.

Again, using the UK as a worked example, the UK National Grid has access to supply and demand from other countries in Europe. For example, there is a 75 km long link between Folkestone and Sangatte in France. This is a 1000 MW bipolar high voltage DC cable, capable of flows in both directions.

To give some idea of the variations in demand, in 2019, day averages have gone from a minimum of 7.67 GW to a maximum of 48.815 GW, with an average of 31.19 GW.

On 22 November 2019, the demand was 44 GW, of which 24% came from renewables. Of this, 16% came from wind and 7% from biomass, with a further 1% each from pumped hydro, hydro, and from all other sources of renewables. At that time of day and in winter it was dark, and hence no contribution from solar. In terms of CO_2, the UK National Grid output was 13,569 tonnes per hour. This was dominated by combined cycle gas turbines at 73%, with a further 10% from coal-fired stations. CO_2 from renewables was down at 2%. Replacing the combined cycle gas turbine generators with a combination of renewables plus modern-day nuclear, plus energy storage, would make a huge impact on UK CO_2. For nuclear power to be included, we must achieve safety, dealing convincingly with the fear factor, and reduce cost. As noted earlier, safe commercial airline travel illuminates the secure pathways available to achieve this, but it is challenging. The fear factor was borne out of large-scale failures that modern design, building, installing, and operating procedures can eliminate. We know how to fix all of these, but every part of the nuclear power generators must work to best practice.

There is an important lesson for climate change. We must not let the fear factor, now expressed by mass rallies, demonstrations, and demonization of old fossil fuel-dependent industries, prevent getting positive actions done.

How might a clean grid look?

Imagine all electricity must come from renewable sources, exclusively from wind turbines and solar panels, we could be in real trouble, within hours. There would be loss of life. There would be days when everything stopped, as the wind dropped, or was too high for safe operation, and it is dark outside. The challenge is one of securing enough electricity supply to meet demand, 24/7. The solution takes the form of optimized energy storage that can be called upon when generation from renewable sources falls short of demand, plus a stable power generator covering much of the load.

Imagine a new grid. Inputs come from renewable wind, renewable solar, and low-cost multi-sourced micro-nuclear in bundles suited to the balance of demand. Nuclear has a level of modulation, but renewables can deliver whatever nature supplies, and we can apply a brake. This grid has three sources of storage energy – batteries, high energy liquid fuel, and hydrogen.

We have batteries for fast response. We can demand an instant response up and down. There may be big industrial batteries, and there will be some domestic batteries hanging on the wall or sitting in the family car outside. For longer periods of renewable power outage, we need either a high-energy liquid fuel or hydrogen, or both.

Added measures for supply/demand matching include switching off power safely until required again or selling excess back to the grid. Conversely, when there is a shortfall, we can use stored energy or import from the grid at the going price. Once we get to a hydrogen economy with few cars and vans using combustion engines, it will be batteries and hydrogen that dominates the energy storage scene.

Buying and selling energy can be expected to remain a big part of the supply/demand balance. Currently, we use the modulation of combined cycle gas turbine-driven generators, but in time they will perhaps be replaced by multi-micro-nuclear! Much work is needed to verify if the potential of micro-modular nuclear can be made safe and cheap enough to offer a practical solution. As we discussed in Chapter 7, critical to its success is a large and secure launch customer base. If that cannot be achieved then nuclear power may be too expensive forever, unless nuclear fusion gets the breakthrough we all want.

Considerable progress has been shown by Kubagen in reducing the size and cost to store hydrogen. This would appear to suit cars and vans. Further progress is needed for long-range hydrogen-powered commercial aviation. For this most demanding of applications, liquid hydrogen appears to be the front runner. The FlyZero programme (www.flyzero.org) will explore this. H2Accelerate[1] is another collaborative programme we'll hear more about in Chapter 9. The programme explores the early introduction of liquid hydrogen-fuelled heavy trucks, with the aim of matching the price and cost of ownership of big diesel trucks, while achieving net-zero emissions. Note the degree of common purpose between commercial aviation and heavy trucks. Hydrogen fuel cells plus batteries look to be a real contender for truck applications, provided we can reduce the volume and pressure in the hydrogen tanks. Liquid hydrogen-powered trucks appear to make the best case. Linde, Daimler Volvo, and Iveco, together with leaders in the fuel supply line, are taking liquid hydrogen seriously, as the means by which they expect to match the purchase price, cost of ownership, and revenue-earning power of diesel trucks. The fuel cell takes care of the long hours at cruise and keeps the battery topped up ready for the next high demand, such as acceleration and hill climbs. When restricted to known practice, using hydrogen for any form of transport demands the compression of hydrogen to 700 bar pressure. As noted elsewhere, the tank is big and very heavy if high leakage is to be avoided. There are some rail operations where electrification using power from the grid is too expensive, hydrogen-fuelled fuel cells plus a modest battery appears to make sense. Here, the hydrogen might be used at a low pressure, marginally above ambient pressure. The tankage requires a volume about 600 times greater than the volume of diesel fuel for the same power at the same range. With the use of hydrogen storage held in manganese hydride, at an intermediate pressure of 100+ bar, we could expect to achieve a realistic volume for cars. The task is eased further in that fuel cells deliver a higher efficiency than the best diesels, thus increasing the range for the same fuel energy content.

For seasonal storage, advanced sustainable carbon-neutral fuels will be used. These must have a high power density per unit weight and per unit volume. Aviation is the most critical, but it remains important for all applications including those on the ground. Space and the ability to support heavy weights have high costs. Direct Air Capture of CO_2 (as discussed in Chapter 5), as the basis for a liquid fuel of high energy density, is a strong contender, and we know how to do that but not yet how to get it to the global mass market. We will need large numbers of direct air capture units placed around the world, close to population centres.

Hydrogen storage in a more transportable form may be achieved using a patented transitional metal-based Kubas hydrogen sponge to give a projected four times the volumetric density of 700 bar storage tanks at five times lower cost.

Summary

As we've seen throughout the last four chapters, the creation of sustainable clean energy requires a multitude of working parts.

We need the ability to generate in a way that minimizes (or eliminates) emissions. Renewables like wind and solar are clearly core to this but we must also be open-minded to the role that small, even micro-modular nuclear power might add to the mix to help smooth out the supply and manage the demand as environmental conditions change.

Storage is an essential component of the solution – if we can't store energy, we will lose generated excess supply and have nothing for when demand peaks. Similarly, without storage, it makes the global transportation of energy difficult if not impossible.

Finally, without a clean grid, none of this matters. The only way we can replace fossil fuels with renewables is to create the ability to bring the energy from places with excess supply to places with excess demand. Mass-produced small modular nuclear has yet to be proven, but the potential deserves proper study by those within the industry plus those who know how to get the best quality through high volume.

We're going to need a lot of help to bring all the components together. We will need governments to create the right regulation and market conditions to facilitate the acceleration of the creation of the infrastructure to generate, store, and transport energy. We will need citizens to play their part, not just in helping to consume less energy but also to be more vocal and engaged to provide the necessary demand that drives priorities and economies. And, of course, we're going to need some really bright design engineering thinking to help us navigate the tricky path ahead.

Note

1 Daimler (2020), *H2Accelerate – New Collaboration for Zero Emission Hydrogen Trucking at Mass-Market Scale.* Available at https://media.daimler.com/marsMediaSite/en/instance/ko/H2Accelerate--new-collaboration-for-zero-emission-hydrogen-trucking-at-mass-market-scale.xhtml?oid=48445607 (Accessed 7 March 2021).

Part 3

TRANSPORT

Critical actions

- Travel and transport less – move data not people
- Utilize existing transportation with a long service life
- Make electric cars the norm
- Long-range commercial aviation and other transportation must have a sustainable liquid fuel

DOI: 10.4324/9781003193470-11

9

CARS

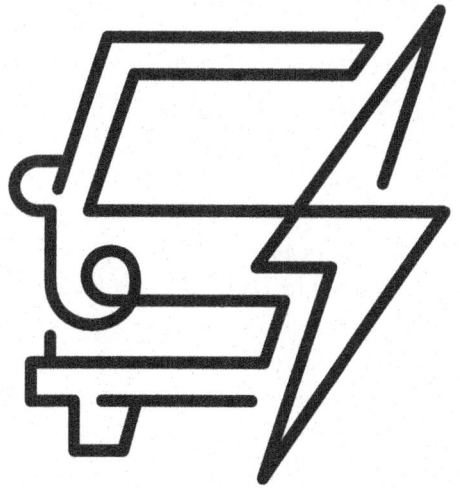

DOI: 10.4324/9781003193470-12

When I was born, the family transport was an AJS motorcycle with a sidecar. Definitely not good for social interaction on a long journey, but my dad had noted that a Morgan three-wheel car could easily outpace the motorbike and sidecar, more than that, there was a family version with four seats. The basic three-wheeler was a masterpiece of design. If the car did not need it then it was absent. Henry Frederick Stanley Morgan founded his company in 1910. His three-wheel car was rich in innovation with its independent and well-damped front suspension. The v-twin engine exhaust gases were silenced before exiting through the tubular frame on the model we had as a family car from 1936 until 1954.

Each year before the long run from Coventry to Greater Manchester, the engine was removed and taken on a wooden trolley to my dad's workshop where the engine was stripped, cleaned, and reassembled before refitment and road test to verify the complete absence of problems before it was looked at next year. I was always there to help with reseating the valves, checking carburettor jets for blockage, and removing all coke deposits.

Mark Twain once said, "History never repeats, but it often rhymes". As I write my book, it's wonderful to see the poetry of Morgan's design principles shine through in the incredible new designs coming through from the likes of Electra Meccanica and Arcimoto and others. There are exciting times ahead.

In the minds of most people, cars are seen to be the worst offenders when it comes to GHG emissions, but there are strong challengers for that honour. What we eat and the homes where most people live are strong contenders for the worst offenders' title. Based upon all the data I have seen and used in my calculations, the data suggest that intensive animal farming claims the prize for worst Climate Change Offender, but cars claim a place somewhere on the "winners" podium for worst offenders! Heavy goods vehicles (HGVs) are also big contributors to greenhouse gases. This has been recognized by truckers. A common-interest alliance called H2Accelerate was formed in December 2020 with the aim of achieving nationwide introduction within a decade.[1] This alliance is backed by Daimler Truck Iveco, OMV, Shell, and the Volvo Group. The aim is to match the purchase price and operating costs of diesel but using hydrogen as the fuel. To retain the payload capability within truck weight limits, the hydrogen will need to be in liquid form. This leads to a shared interest in long-range commercial aviation.

Cars, buses, and trucks present two problems. First, a range of vehicles with zero emissions is needed for entry into service as early as possible. Second, the world will shortly contain two billion conventional cars, with huge numbers of buses and trucks, many of which have a long useful service

Figure 9.1 Daimler's vision of a liquid hydrogen-powered truck as part of their involvement in the H2Accelerate programme.

life ahead. For these, we need the best possible fuel in terms of emissions until all of them can be taken out of service and recycled. Many modern cars have a flexible fuel capability, meaning they could run on a carbon-neutral liquid fuel. Other cars can be adapted, but many more must be removed from service as quickly as possible, while globally flexible fuel cars can benefit from fuels like those from our existing stock of modern jet planes.

The world has around 1.3 billion cars.[2] As of 2019, 35 million are in the UK.[3] On average, a UK car does about 12,000 km/year[4] releasing about 151 g CO_2 per km (using an average to allow for average emissions per year for an average age of car).[5] Doing the maths based on UK numbers, our cars add 63 million tonnes. UK CO_2 output is 370 million tonnes. Hence, UK cars contribute c. 17% to UK gross CO_2. If you extrapolate this to a global level (which is admittedly crude), the world contribution to climate change for driving cars amounts to $0.151 \times 12 \times 1.3 = 2.35$ billion tonnes of CO_2. In 2019, the world annual output of CO_2 was about 36 billion tonnes, so world cars are adding to the problem at a rate of 6.7% per year.

Many are told that battery-electric cars are the way forwards, but is that true? Every battery-electric car does help, but the problem is that first-generation electric cars are too expensive for the mass market. Moreover, battery-electric cars must be manufactured. This incurs a high penalty of between three and eight tonnes of CO_2 per car (depending on battery capacity and the current state of battery technology and production methods).[6] This will reduce as a higher concentration of renewable energy is used. Moreover, many electric car users will find that an electric car with a lower range meets most of their needs. There is a need for electric hybrid cars, in which the cruise power comes from a fuel cell, with a smaller battery to give a boost for acceleration. The combination of high price, even with a generous subsidy, plus the hassle of charging and range anxiety restricts the uptake of battery-electric cars. In turn, this limits the climate change benefit. Excellent progress continues to be made in increasing the range of lithium battery-powered cars, but the bigger batteries cost more such that only the affluent can afford them. This reduces uptake, which reduces the benefit for climate change. We need a better form of electric car, one that can reach the mass market. Far more effective is a change in fuelling as many cars as possible, with a synthetic carbon-neutral fuel. Coupled with much-reduced travel, the benefit could exceed 10% of the human annual output of CO_2. Ultimately, everyone must transition to pure battery power, or if range between refills and fast refill times are needed, then a fuel cell plus battery hybrid becomes a desirable choice. Confirmation that hydrogen fuel cells can run in reverse adds greatly to the attraction of electric fuel cell battery hybrid cars. Many hydrogen fuel cell-powered cars may rarely need to visit a roadside refilling station, as most needs can be met by recharging at home overnight, ideally using renewable electricity. The UK has a discounted lower price per kWh, making overnight charging within the discounted hours attractive. Not every country has reduced price electricity for times when demand is below supply.

We must also move towards cars that are lighter in weight and lower in drag. All electric and hydrogen cars are good but unless they are affordable and able to meet customer needs there cannot be enough of them to help in avoiding a climate disaster. More efficient drive trains will reduce the aggregate power needed to be generated from renewable sources, it can then be stored and transported to where it is to be used.

By separating out cruise power demand from the need for short but sharp acceleration, we no longer need a powerful engine that is far too big for cruising at legally allowed speeds.

I remember thinking about this during my last days at school. At that time, I could only think of a supercharger booster! Now we have powerful lithium batteries and supercapacitors to take care of the acceleration. These are recharged en route when available cruise power has excess. By downsizing the car, splitting cruise from acceleration, we can reduce emissions.

There is scope for another class of vehicle with much-reduced performance and power that sits in the quadricycle class with performance restricted to that at which a 14-year-old can drive without a driving licence. This gives teenagers new freedom and saves parents running a taxi service with higher emissions. This could be a powerful way of reducing emissions while retaining personal mobility for many of those needing that, as well as creating a whole new industry.

We also need to reduce the number of cars by using on-demand services, and by reducing the need for many of the journeys we take. As we will show later, it is possible to produce ultralight cars with low drag that are well suited to the transport of single or twin occupants. At this time, they would be classed as motorcycles, despite many of them having all the in-car amenities of a Range Rover, a Mercedes saloon, or a Cadillac from the United States. The market entry model would be affordable for the consumers in Level 2 and Level 3 incomes, while those with higher spending power would buy the most lavishly equipped version. It could be used as a strong signal of caring for others and caring for the planet we all share.

Long-range trucks powered by fuel cells running on liquid hydrogen will match the range with a full payload of the diesel equivalent. It follows that hydrogen can easily meet the range requirements for fuel cell cars even from the most demanding customers. Few fuel cell cars will need liquid hydrogen. Compact storage at reasonable pressures will do a great job. For many with off-road parking, charging at home may be the only fuelling station needed as the range potential of hydrogen-powered fuel cell cars can have a usefully long range between refuelling.

Electric hybrid cars

Electric cars powered by a combination of fuel cells and batteries have the potential to be cheaper and cleaner, with good performance, and more range than battery-only cars. Battery cars are being pushed hard as the decade of the 2020s begins. We can understand why, as it is "the only game in town", but we really need maximum effort to get the ultra-low emission cars people will buy in mass-market volumes. This needs an appropriate hydrogen fuel distribution network but note the massive impact of being able to generate the hydrogen at home overnight. The UK has very efficient fuel cell technology. Hydrogen fuel cells deliver electricity at over 60%. An electric driveline means that power to the driving wheels could be nearly twice as efficient as that of a diesel, with all its parasitic losses for cooling, gears, and heat management. Batteries are big, heavy, and expensive when used as the sole source of power. A fuel cell, sized for legal maximum cruise speed on motorways, plus a much smaller battery to take care of acceleration and long steep hills, has the potential to reduce the purchase price of a car and avoids the need for battery charging from a ground-based source. Hydrogen fuel cells cars need to reach high production volumes in order to get to the right price point.

Flexible fuel cars, vans, and trucks can use fuels such as E85. France has started to convert petrol stations to sell a liquid fuel known as E85. Petrol and diesel fuel are taxed as now, while E85 in France is tax free. E85 is 85% ethanol blended with 15% petrol. It burns cleaner and many cars can already use it. The ethanol can be made using CO_2 captured directly from the atmosphere, then processed using renewable energy. Other cars may need some engine management system changes. Solid oxide fuel cell would happily make electricity to propel an electric car at a far higher efficiency than any petrol or diesel engine. For the battery-only powered car charging from the grid incurs the release of CO_2 until such a time when fully renewable/stored electricity is available. A typical battery car needs 15 kWh per 100 km for which the UK grid of 2019 releases over 4 kg of CO_2. The message is clear. Electric cars are good for local air quality and climate change, but until they dominate the mass market their help in slowing climate change will be minimal. For this to happen, electric cars must become much more affordable, and they must become easier to live with. The alternative is a much reduced electric car range, with increased use of public transport.

Published data suggests that on average electric cars need 15 kWh/100 km (based on an example of a Tesla Model 3 Sporting with a 76 kWh battery and a claimed range of 511 km).[7] If all UK cars were replaced by electric cars, we need 35 million electric cars. The aggregate UK demand for electricity is 63 million MWh. The CO_2 generated by the national grid is 270 g/kWh

(From GridWatch UK data showing a 2020 average of 7.918 kt/h CO_2 created from 29.653 GW of power generated. This equates to 267 g/kWh).[8] The electric car incurs 17 million tonnes of CO_2 for all UK cars/year. As we've seen earlier, with the average petrol or diesel car puffing out 151 g/km, the total output for the same number of cars with the same annual km, the output of CO_2 is 63 million tonnes. The electric car is the clear winner by a huge margin. The electric car is 73% lower in CO_2. However, there is a massive task in getting everyone to replace their conventional car with one powered by batteries charged from the National Grid. The resistance will increase when prospective buyers realize that electric cars with fuel cells plus a smaller battery will do a better job. If 10% of the world's cars changed to battery-only powered electric cars, it would drop CO_2 by 1.1% of the world's total. The real priority is to reduce travel by cars, followed by upgrading all remaining cars to run on a clean fuel, preferably hydrogen.

In many countries, there is a need for an urban runabout car suitable for taking up to four people for local shopping or other duties needing only a short range and a modest top speed.

Consider the example of the i-MiEV from Mitsubishi Motors Corporation (and we will explore other examples of this approach in more detail in the next section). This is a small four-seat car capable of 130 km/h. It requires 10 kWh to cover 100 km. For 12,000 km/year, the car needs 1200 kWh/year. If the car is charged from the UK National grid, the CO_2 per car per year is 270 g/kWh, therefore, each electric car produces 324 kg CO+/year. If all 35 million cars were like this example, then the CO_2 drops from 63 million to 11.3 million tonnes, a drop of 82%. At this level, UK gross CO_2 from cars reduces from 17% to 3%.

However, it will take far too long to replace all cars in the UK with small electric cars powered by batteries. Department of Transport data suggests there were around 277,000 electric cars in 2019 (both plug-in hybrid and battery only). This represents a market penetration of 0.8% (that number drops to 0.26% for battery-only cars), so climate change reduction is almost zero.

There are 35 million UK cars, still increasing in number, which must be run on a much cleaner, carbon-neutral fuel until it is time to remove them. We must reduce our need and desire for travel. Technology makes it realistic to reduce travel drastically. Based on experience both in and out of the pandemic, 80% seems a good target for reduced travel.

When we must travel, the process needs to be with zero emissions, or as close to that, as is achievable. As such, changing the fuelling to one that is close to carbon-neutral may have a much more powerful effect than the slow shift to expensive battery-only cars. Already there are commercial fuel providers and distributers able to provide E85 petrol ethanol mix. E85 contains 85% ethanol, blended with 15% petrol. Many cars will run well on this fuel.

Although the specific savings depend on the source of the ethanol (figures referenced from the Argonne National Laboratory [United States] show a range of lifecycle GHG emission reduction of 19–48%, 40–62%, 90–103%, 77–97%, and 101–115% for corn, sugarcane, corn stover, switchgrass, and miscanthus sources, respectively).[9] Taking a mean of the median of these ranges, we get 75% as a rough guide to the overall average available reduction in GHG emission from a range of biological sources.

If all UK cars were to use a carbon-neutral version of E85 for all journeys, the CO_2 saving is over 47 million tonnes/year, which is nearly 13% of the UK's annual emissions of CO_2.

E85 is available now, but only in a few outlets. In France, the government has installed many E85 pumps at regular petrol refuelling retail outlets, but only in a few key cities. It is a start, and it points to a useful and realistic way forwards. In addition, the French Government has removed the tax on E85 while retaining tax on sales of petrol and diesel.

This action helps now, even though the ethanol does not yet come from the cleanest source. Further reductions in emissions require the ethanol to come from the direct air capture of CO_2, followed by a clean electrically powered process that converts CO_2 plus water into liquid ethanol fuel. If the ethanol comes from crops grown on good land, the gain is reduced. Some is from farming, but the biggest adverse effect would come from importing food we could have grown on the land now growing feedstock for ethanol production. However, more intense farming methods can greatly increase yields per unit area of land. In winter when ambient temperatures drop, the ethanol content may have to be reduced. If the synthetic production of ethanol used national grid power, instead of clean solar power at the grid's present level of CO_2/kWh, the gain is reduced. It's another reason why we must upgrade UK electrical power supply to be greener as fast as possible. Ethanol has a bunch of undesirable characteristics for which solutions are yet to be secured for mass-market sales. Ethanol is easy to ignite, causing fires. Paradoxically, it is also hygroscopic, which means it seeks water that leads to problems with ice blocking fuel lines. None of these are insurmountable, but it will take time to get everything right for the mass market. Electric cars also have a big list of characteristics for which solutions are needed before mass-market penetration can be achieved. Liquid fuels have a high energy density per unit weight and volume.

Ethanol from carbon capture using renewable energy is a strong contender for reducing climate change damage for personal travel. The most attractive power source for a car is one using a carbon-neutral fuel or hydrogen in a fuel cell to make electricity that feeds the electric motors and battery. Fuel cells convert the energy from liquid ethanol into electricity, with an energy efficiency of 60%, or more, plus some useful heat that can be used as heat or for refrigeration. Some of this heat could be used to keep the battery within the right operating temperature range. This is close to average day temperatures. By following this route of fuel cell plus battery, cars could be carbon neutral with no range restriction. Such cars have no need for a battery-charging cable or a source of electricity from which charging can take place.

Hydrogen is a cleaner fuel, and it widens the choice of fuel cells to embrace polymer electrolyte membrane fuel cells (PEMs). Hydrogen has a high energy density per unit weight. It is 2.8 times more energy dense per kg, but fuel tank weight and volume erode that advantage. Hydrogen energy density per litre is very low.

With long-range trucks needing to get to zero emissions quickly through the use of liquid hydrogen, there should be scope for mutually supportive work on solving shared design challenges.

Ultralight runabout electric vehicles

As the electric car industry develops, things will improve with respect to eliminating the negative aspects. Competition has already started. Second-generation electric cars are starting to overtake petrol cars in terms of cost of ownership. Moreover, there is a huge potential market for very affordable electric cars, including minimalist vehicles, such as urban-only shared quadricycles. While to many this will seem to be fanciful, we need to note that Honda alone has produced 400 million lightweight commuter scooters.[10] This is equivalent in units to 30% of world cars, and that from only one manufacturer. There are more manufactures adding to this total. How big will the market be for something with two seats and weather protection? If 30% of world cars were electric runabouts, the global saving of CO_2 would be around 2% of the world's current annual output of CO_2. Car journeys currently contribute about 2500 Mt CO_2 (7%) to the global CO_2 emission total. Reducing car journeys by 80% brings the contribution down to about 490 Mt CO_2 (1.4%). If the remainder uses E85, clean fuel we could bring this down to somewhere around 122 Mt CO_2 (0.35%), which is a whopping 95% reduction on where we are now.

If we could do similar things for delivery trucks and vans, we can reduce greenhouse gas emissions from 6300 Mt CO_2 (18%) to 312 Mt CO_2 (0.88%).

Admittedly 80% in road transport is a big reduction, but reduced travel made possible by increased use of technology is a powerful way of reducing greenhouse gases and as we've learnt from the pandemic, we know now we can travel a lot less without a significant reduction in outcomes or experiences.

For those workers still needing to travel to work, there is a growing need for a streamlined, lightweight form of electrically powered personal transportation with only one or perhaps two seats. This can be thought of as a motorcycle with weather protection. This vehicle may take many forms, with three or four wheels, and sufficient performance to run smoothly within heavy traffic. At the top end, the vehicle would look very smart, and have a top speed of 80 mph, with a lively acceleration to keep pace with the other vehicles joining a fast stream of traffic. Top-end versions would have air conditioning and a high level of in-car entertainment. The lowest range would be the equivalent of a pedal-assist electric bike. Top speed would be restricted to the maximum allowed for a 14-year-old teenager. This is about 15 mph. It gives teenagers freedom of movement for journeys to school, universities, or to run errands for those in need. It is important to note that a single-seat motor vehicle has the potential to be completely resistant to the transmission of a virus posing a health hazard. Clearly, masses of such vehicles could overload our road networks, but with many people working from home, personal transportation might well be a good way forwards.

Citroen has often been a leader in ultralight and basic cars. In 2019, the Citroen Ami was first shown. In production form, it is expected to be a mere 2.5 metres long, urban two-seater that can be driven without a licence. With a top speed of 45 km/h and a full range charge of 70 km, it is aimed at city car-share fleets and cash buyers. It is a comfier alternative to a scooter or moped.

Electra Meccanica of Canada is trialling Solo, a small battery-electric single-seater car with full weather protection, air conditioning, and good performance with a price at about half that of the average new car purchased in the United States (Figure 9.2). The new bubble car looks smart, has a top speed of 80 mph and a range of 100 miles. The trials are in Los Angeles, California.

Similarly, a little further south of the border in Oregon, Arcimoto has released a dual-occupancy commuter vehicle, the "FUV" (fun utility vehicle; Figure 9.3).

Both cars have a low centre of gravity and crash protection. Moreover, they are as tall as many regular cars to help other drivers see them in mixed traffic. We can expect to see more electric vehicles with increased driver appeal at prices that are more affordable out of earned income for most workers.

My sense is that both of these designs are just the beginning of a lot more innovation to come and it will be interesting to see if it leads to a large growth in personal electric transport.

Figure 9.2 Electra Meccanica Solo – single-occupancy commuter vehicle.[11]

Figure 9.3 Arcimoto FUV – dual-occupancy urban commuter vehicle.[12]

For those countries with inadequate supplies of electricity, there could be a petrol equivalent in which a tiny petrol engine drives a small generator to power the electric propulsion motor and charge a small battery that boosts acceleration. The benefit in emissions would only be slightly less. Since the fuel demand for the ultralight urban vehicle is incredibly low, many communities will choose to fuel them with ethanol produced locally from plants grown close by. Where renewable power is readily available, the local production of hydrogen becomes realistic. The tiny combustion engine will run on hydrogen and not produce any significant greenhouse gas emissions.

I hear many people screaming at me. If all our cars are electric why bother with these toys? Climate change is global, and most of the world cannot afford lovely electric cars with a decent range. The more expensive electric cars are welcome and will help climate change, but they may not get the world market share needed to benefit climate change by the amount required from world cars. For many decades, big cities like Shanghai have used micro vehicles well suited, but not restricted, to the many very narrow streets and passages.

Despite the trend to on-demand taxi services, there are still many young people wanting to learn to drive, and who need to drive to get the work for which they aspire. It is important for these people to have the opportunity to drive electric cars from the start of their life of driving. The most promising solution uses a flex fuel powered fuel cell, sized for maximum legal cruise speed, plus a battery for speed increases and energy recovery during braking. The design of the car needs optimization.

There needs to be a clean source of carbon-neutral fuel for replenishing the contents of the small fuel tank. If the equivalent petrol car does 50 mpg, the fuel cell electric hybrid could double that, provided it is optimized.

Success is often measured by the car you drive. Climate change says that is an unacceptable luxury. We need to be judged by our care for our climate and our health. The original Mini attracted support from the rich and famous while it made an ideal first car for many young families. It helped that Mini won the famous Monte Carlo rally in January 1964.

Existing cars

We must take action to reduce the damage to the climate. Everyone that can, must reduce mileage and keep cars longer. We must retire the worst offenders. Nations might adopt a system based on the one used in Singapore, involving a permit of ownership. This certificate might be used to severely limit ownership and use of the worst producers of greenhouse gases. We need to fix most if not all 1.3 billion of them.

The priority must be to reduce the number of cars, and the miles travelled by cars. The best solution from the climate change point of view, for those still required, would be to run normal petrol engines on hydrogen by modifying the fuel management system to make that possible. BMW has paved the way, and other leading manufacturers have similar technology, but until we have practical and cost-effective hydrogen storage with a realistic supply infrastructure, it falls well short of a mass-market solution. The emergence of a four times higher volume density energy storage at five times a lower price using hydride storage could change that. There is still much work to do to get to a global mass-market scale, so a liquid fuel that is carbon neutral may have to be a compromise solution for many owners unable to convert to hydrogen-fuelled new vehicles with the better fuel cell plus battery technology.

All renewable forms of energy need energy storage, some of which may be offered at a low price, especially at times when supply exceeds demand. This might make hydrogen widely available at a competitive price.

The evidence shows that a small fuel cell sized for the legal maximum speed on freeways/motorways, plus a small lithium battery to take care of acceleration and long hill climbs, is a realistic way forwards that many people can afford and will enjoy driving, especially as it may be practical for flat dwellers living in busy urban areas. Experts in fuel cells and fuelling infrastructure will decide on the choice of fuel cell, with the choice being between proton-exchange membrane (PEM) and solid oxide (SOFC).

The powerplant of a car sets the design

The choice of power plant for a car has a strong influence on car design, manufacture, and driving considerations. The car must be optimized for all these constraints to get the best ride quality and the best functionality at the lowest price. Petrol and diesel cars find that the power unit has a strong influence height of the scuttle (the part of a car's bodywork between the windscreen and the bonnet/hood), and hence on the drag of the car. In turn, this increases the cruise power demand. The engine also influences the weight of the car. This demands more power for proper levels of acceleration. Fuel cell-powered cars provide the opportunity to reoptimize car design and reduce the demand for high levels of power while retaining all the qualities of space, acceleration, and comfort of ride.

Some citizens will be pleased with cars that can only be used as local runabouts. This is especially true when the need for cars is reduced through technology and the removal of much daily commuting. Less frequent long-distance travel can be hired, when the need arises, at a much-reduced cost than ownership of a general-purpose car.

In the transition from fossil fuel to renewables via energy storage, the treasury has lost several major sources of revenue needed to provide essential services from the government, so we can expect to see new forms of tax, such as road pricing. A tax policy for quickening the pace towards zero emissions is needed, but some provision will be essential for low-paid workers with a dependency upon transport.

Note that such a tax reduces the attractiveness of the electric cars needed to reduce greenhouse gases from road transport. Every change we make on the path towards sustainable life for humans on Earth impacts almost everything else. This shows why we need a total system approach to ensure that all the good individual changes result in a new workable solution that is good for everyone.

Notes

1 Daimler (2020), *H2Accelerate – New Collaboration for Zero Emission Hydrogen Trucking at Mass-Market Scale.* Available at https://media.daimle r.com/marsMediaSite/en/instance/ko/H2Accelerate--new-collaboration -for-zero-emission-hydrogen-trucking-at-mass-market-scale.xhtml?oid=4 8445607 (Accessed 7 March 2021).

2 Wards Intelligence (2017), *World Vehicle Population Rose 4.6% in 2016.* Available at https://wardsintelligence.informa.com/articles/2017/10/17/wo rld-vehicle-population-rose-46-in-2016 (Accessed 7 March 2021).

3 Society of Motor Manufacturers and Traders (2020), *Motorparc Data 2019.* Available at https://media.smmt.co.uk/vehicle-ownership-in-the-uk-surpas ses-40-million/ (Accessed 7 March 2021).

4 GOV.UK (2020), *Vehicle Mileage and Occupancy.* Available at https://ww w.gov.uk/government/statistical-data-sets/nts09-vehicle-mileage-and-oc cupancy#car-mileage (Accessed 7 March 2021).

5 Department for Transport (UK) (2020), *Vehicle Statistics/Cars VEH0206: Licensed Cards by CO2 Emission and VED Band: Great Britain and United Kingdom.* Available at https://assets.publishing.service.gov.uk/governmen t/uploads/system/uploads/attachment_data/file/882296/veh0206.ods (Accessed 10 March 2021).

6 Swedish Environmental Research Institute (2019), *New Report on Climate Impact of Electric Car Batteries.* Available at https://www.ivl.se/english/ivl/ topmenu/press/news-and-press-releases/press-releases/2019-12-04-new-r eport-on-climate-impact-of-electric-car-batteries.html (Accessed 11 March 2021).

7 Tesla (2021), *Tesla Model 3.* Available at https://www.tesla.com/model3 (Accessed 11 March 2021).

8 GridWatch UK (2021), *GB Fuel Type Power Generation Production.* Available at https://gridwatch.co.uk/ (Accessed 11 March 2021).

9 Wang, M., Han, J., Dunn, J., Cai, H. and Elgowainy, A. (2012), Well-to-wheels energy use and greenhouse gas emissions of ethanol from corn, sugarcane

and cellulosic biomass for US use. *Environmental Research Letters*, 7(4), p.045905. Available at https://iopscience.iop.org/article/10.1088/1748-9326/7/4/045905/pdf (Accessed 11 March 2021).

10 Honda Europe (2019), *Honda is Celebrating the Production of 400 Million Motorcycles*. Available at https://hondanews.eu/eu/lt/motorcycles/media/pressreleases/199277/honda-is-celebrating-the-production-of-400-million-motorcycles (Accessed 11 March 2021).

11 Electra Meccanica (2021), *Solo – Electrify Your Commute*. Available at https://electrameccanica.com/solo/ (Accessed 11 March 2021).

12 Arcimoto (2021), *FUV – The Fun Utility Vehicle*. Available at https://www.arcimoto.com/fuv (Accessed 11 March 2021).

10

COMMERCIAL AVIATION

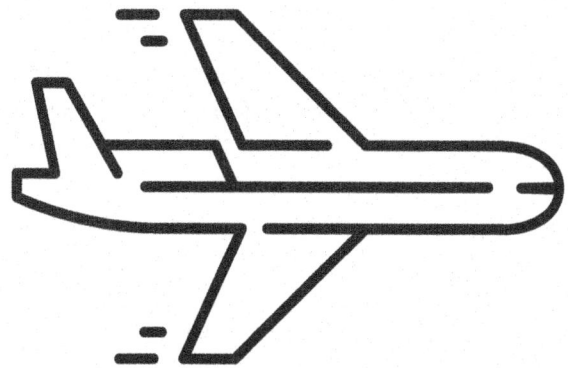

DOI: 10.4324/9781003193470-13

My love of commercial aviation started just after the end of World War 2 when I started to become aware of innovative aeroplane design. This was at a time when I needed to decide what my next step after leaving school might be.

I saw the AW52 Flying Wing and was aware of new gas turbine-powered passenger aircraft starting to enter commercial service. Using my knowledge of designing and making models, plus my view of industries, pointed me towards aeronautical engineering. Imperial College was the most appropriate choice for me because the teaching was more comprehensive than other great universities at that time. Better still, it was in central London where all the engineering leaders came to give talks and mingle with students.

I took the entrance exam for Imperial and gained a place. Imperial College and the learned societies did not disappoint. I met most of the leading designers, and years later when I met them again, they remembered students like me. From university, I gained a graduate apprenticeship with Rolls-Royce, where I soon found a post with the company's advanced projects designer, Geoff Wilde.

Rolls-Royce was brilliant in many ways by challenging its graduate apprentices to solve problems the professionals were finding hard. This worked well for me. More than this, they exposed me to what was going on in the world through business visits, often on my own, to work with innovators in countries far and wide. This was a huge learning experience, first with the defence side, but it was soon clear that commercial aviation was pushing harder, as every improvement gained market and financial rewards. The military could afford to wait. Wars drove their pace!

I feel proud of helping make it possible for millions of workers and their families to live their dream of flying safely to anywhere in the world, tempered by huge waves of shame as this growth in traffic continued to add pace to an emerging climate disaster.

But it's the knowledge that with good design, we can solve many problems. I know we can use the same great talent to innovate and enable commercial aviation to achieve zero emissions, such that in time even more people will be able to fulfil their dreams of foreign travel. The task is exceedingly challenging, but Mother Nature has enabled humans collectively to find and use the right tools. Early work has begun, but a much quicker pace is essential.

We have seen earlier that commercial aviation has not been a major contributor to climate change. However, prior to the coronavirus pandemic, it was set to become one of the big contributors within the next couple of decades. Moreover, there were concerns that some of the damage from jet exhaust emissions were underestimated. We know that flights in excess of 2000 NM dominate the emissions from commercial aviation. We also know that following Brexit, and the accelerated growth in the industrialization of China, India, and the huge populations in Southeast Asia, that long-range trading is set to grow faster than more local trading. Long-range trading requires long-range aircraft in increasing numbers.

Avoiding a climate disaster requires a solution that has zero emissions and there is only one fuel that can make that possible. Liquid hydrogen is that fuel, and that delivers many challenges. However, collectively, there are enough scientists and engineers who understand these challenges to be able to deliver zero-emission aviation. One example of such thinking is again from the team at ZeroAvia (Figure 10.1).

Some companies and research groups have clear plans, with strong actions to pave the way towards liquid hydrogen-powered commercial aviation. For long-range direct burn gas turbines, powering large and very efficient fans will be a dominant candidate. Hydrogen has almost three times the energy/kg than jet fuel, but even as a liquid, it needs four times the volume. Moreover, that volume must be well insulated to minimize fuel boil-off. For the equivalent of a long-range Airbus A350, the fuel volume for the liquid hydrogen is around 500 cubic metres which is double the volume of the baggage hold. This large volume must be found in a way that reduces drag. Happily, the blended wing body (BWB) shape does that. The blended wing body shape looks rather like a Vulcan bomber, but with stretched wing tips to increase its wingspan and increase aerodynamic efficiency. As we'll see later, pictures showing these high-performance designs are known to present challenges for the design engineers, but there are real signs that these challenges are surmountable with realistic engineering. In a possible example, imagine we take the pressurized body of the long-range Airbus

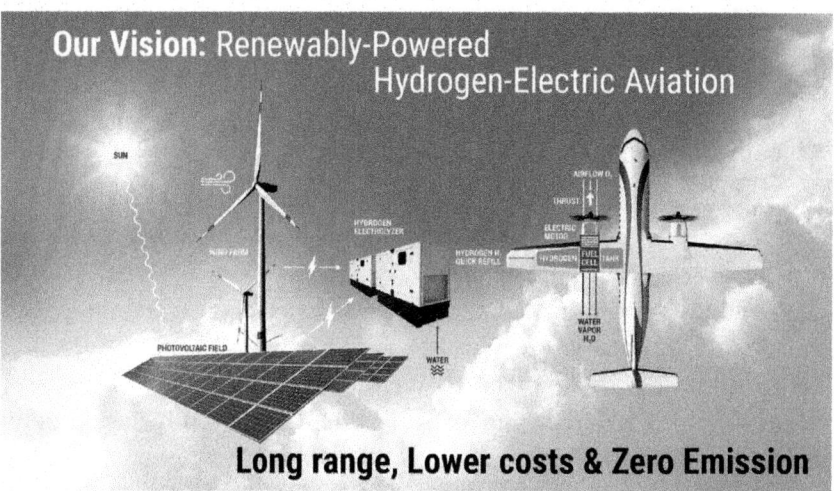

Figure 10.1 ZeroAvia's vision for zero-emission, long-range commercial aviation.[1]

A350 complete with its tail. We then surround this with a blended wing body that is unpressurized, using the side parts to house the liquid hydrogen fuel tanks. We can propel the aircraft with two liquid hydrogen direct burn geared fan engines of very high efficiency placed over the wing. The wing provides some shielding of the noise from the propulsion system to reduce the aircraft noise footprint that is heard by observers on the ground.

There is another choice that might give a better result. We could remove the big gear-driven fans and let the gas turbine drive an electric generator that powers a bigger number of electric-motor-driven propulsion fans. This might give a better overall result. Proper studies and tests will direct the choice.

The one variety of greenhouse gas this solution might retain is that of NO_x. Engine development work on improved combustion might deliver a solution. The absence of carbon in the fuel deals with CO_2, and it might also avoid contrails that may form as ice crystals around very fine particles of carbon in the jet exhaust of a jet-fuelled turbine.

Contrails are formed in the exhaust gases, as they meet the cold air through which the aircraft flies at cruise altitudes. Within the exhaust gas is a high concentration of water vapour that is rapidly cooled to form ice crystals, so we might expect a hydrogen-powered aircraft to be worse. However, the ice crystals need a tiny particle around which they are grown. At ground level, the atmosphere is loaded with fine particles, but it is too warm for the formation of ice crystals. At high altitude, there may be fewer tiny particles, except those produced by the jet engine burning fossil fuel. These come from the carbon in the fuel. Liquid hydrogen may avoid the formation of ice crystals. We need to check this out. There may be a residual problem with sustainable aviation fuel if, as is likely, the carbon content of the fuel remains high. Studies are ongoing to understand the interaction between jet exhaust and the atmosphere. We already know that normal jet exhaust can amplify cloud coverage. Steering the aircraft around such clouds is one option for easing the problem although this results in a small penalty of increased fuel burn.

Present plans aim to make it possible to achieve long-range jets with zero emission by 2035. Avoiding a climate disaster requires a faster pace!

For commercial aviation with shorter ranges, the field of design choices widens to include the replacement of the liquid hydrogen-fuelled gas turbine with a fuel cell. Huge progress has been made in getting a fuel cell that is light enough to fly as the main source of power for propulsion. For distances of 2000 NM or less, it starts to become a real possibility. It solves the NO_x problem. There are two contenders for fuel cells: both can use hydrogen but only the solid oxide fuel cell can run on less than pure hydrogen. The solid oxide fuel cell contains an in-built reformer and could use sustainable aviation fuel in liquid form. Others prefer the PEM style hydrogen fuel cell. Some serious players prefer this form of fuel cell. Experts and product designers, together with those who must provide the infrastructure to deliver the fuel, will need to decide which fuel cell will be the choice, but at this time it looks as if we will see both in widespread service.

Here in the UK, following the successful flights of ZeroAvia's Hydrogen fuel cell powered 6 seat aircraft, the UK government has backed Cranfield to convert a 9-passenger Britten Norman Islander to achieve full certification for commercial service in 2024 between mainland Scotland and the isles located close by. This is expected to be the world's first commercial hydrogen powered commercial service, with the first flight being planned for sometime in 2022.

For commuter aircraft up to 70 seats with lower flight speeds and much shorter ranges of a few hundred miles, the case for fuel cells strengthens. Here the fuel cells will deliver the electrical power to drive two or more propulsion systems. In most cases, these will be electric motor-driven propellers.

To achieve the high-power density aircraft needed to fly economically, commercial aircraft will use high voltage, usually 3000 volts. Such a high voltage needs great respect and care for safety at all times. We also need high currents, and as currents change, power cables can produce large forces meaning that we must choose a routing that minimizes all the whiplash forces, and gives secure control, of the cables. Superconducting wires may be needed for low power loss. This might fit well with aircraft fuelled by liquid hydrogen carried at temperatures below $-253°C$.

Battery-powered aircraft have severely limited range. This may change with more advanced batteries, such as zinc-air, or the use of Superdialectrics supercapacitors. We are still a long way off being able to fly big commercial aircraft for long-range using stored electrical power.

Batteries need to improve by two orders of magnitude to be viable in long-range commercial aviation. However, gas turbine-driven generators powering remote located electric propulsors improves the picture greatly. Such a gas turbine could be fuelled by liquid hydrogen, but that needs large and well-insulated fuel tanks. For small low-speed aircraft, this is easier to provide than it is for the much faster and larger aircraft needed to fly the long routes such as London to Perth in Australia. This explains our interest in the blended wing body aircraft design and derivatives of it.

For shorter ranges still with fewer passengers, we enter the new field of electrically powered aerial taxis. These might have only four seats, but they are most likely to be capable of vertical take-off and landing (VTOL) while being able to cruise at speed of up to 250 mph for distances up to 200 NM. UK aviation company Vertical Aerospace provides a great example with its VA-X4 aircraft, currently scheduled for flight testing in 2021.[2]

We might see VTOL carried forwards to commuter flights with about 19 seats. The arrival of high-power density fuel cells, hydrogen and high-power density electric motors and generators has opened up a whole new world of affordable flight. Vertical flight mode requires a lift/weight ratio of 1.2 compared with the thrust/weight for conventional flight of 0.3, but the duration can be short at around two minutes for both take-off and landing. This means that it makes sense to fit high-power density batteries and/or supercapacitors to provide the extra boost. The integration of the two forms of propulsion points strongly to electrically driven fans or propellers, some of which are shut down for cruise. Those that are shut down will be placed in a low drag mode. Some use stowable propellers.

Figure 10.2 Vertical Aerospace's VA-X4.

For flights covering more than a few miles, a wing is needed as this gives the right lift for the lowest fuel consumption. The propulsors must tilt and many also prefer to tilt the wing. Compared with a conventional helicopter, these electrically powered thrusters greatly reduce cost. In turn, this grows the size of and numbers of markets. With hydrogen, as the fuel emissions are zero, such vehicles can operate in urban settings.

Small aircraft benefit from the "cube square law". Double the size and the weight goes up eight times, but the lift goes up by only four times. When we think of aerial delivery drones, they will be very small compared with a passenger-carrying equivalent. This means that small drones are capable of useful commercial service when fuelled by compressed hydrogen at pressures close to that of a scuba divers air tank. Smaller surveillance drones that can swoop into enclosed areas could have useful endurance while being hard to detect. Small Earth-bound delivery drones can also be well served by hydrogen as a fuel.

Aircraft and propulsion design factors

Even liquid hydrogen at temperatures below −253°C is too bulky to fly in a conventional commercial airliner, requiring four times the volume of jet fuel. It needs the extra internal volume of the blended wing body design of an aircraft. Happily, this style of aircraft has other advantages. It offers less drag, a reduced wingspan, and more flexibility for seating arrangements. Work has started to gather the technology to take this into mainstream commercial aviation. Airbus has made a solid case for fuelling aircraft with hydrogen. For the longest flights, liquid hydrogen is needed together with a new configuration having more volume available for the fuel. These aircraft may have to burn the hydrogen directly in a gas turbine. New technology indicates that electrically driven propulsion fans might give the best overall economics. For shorter-range aircraft, the fuel cell has advantages.

An aircraft such as the long-range Airbus A350 weighs about 300 tonnes at take-off and carries around 100 tonnes of jet fuel. To provide the same energy content, we need 36 tonnes of liquid hydrogen, giving a fuel weight saving of 64 tonnes per long flight, some of which is needed for the fuel tanks. As we'll see shortly, the aircraft will look different with a blended wing body. My numbers are simplistic, but they do indicate a means by which we could fly from London to Perth in Australia with almost zero emissions. It is worth noting that liquid hydrogen needs a total tank volume of about 500 cubic metres. The tanks need near-perfect insulation. Compartmentation is required to deliver all the roles played by the fuel on a commercial jet. Fuel is used to achieve the right balance needed for safe and efficient flight. This is normal. We need the centre of gravity to sit directly below the centre of lift.

My own view is that commercial aviation will have to embrace sustainable liquid aviation fuel, which is a blend of carbon-neutral liquid fuels blended with aviation kerosene, but we do need to quicken the pace towards the replacement of conventional jets with blended wing body shapes that can provide the huge volume needed for liquid hydrogen fuel for efficient turbofan engine propulsion.

Aircraft and engine designers are raising aerodynamic efficiency, reducing fuel consumption, and reducing engine and aircraft structural weight. Each of these improvements reduces emissions. The electric hybrid powering of aircraft gives new freedoms, helping achieve aircraft that need lower emissions per passenger mile. However, a high-energy liquid fuel remains the only means of flying long-range, existing commercial jets, but we do know how to get closer to a carbon-neutral fuel. As a direct consequence of reducing commercial airline emissions per passenger mile, we have made aircraft much more accessible to the ordinary person. In turn, this has encouraged the commercial airline business to grow rapidly, leading to commercial aviation causing a big increase in greenhouse gases. For commercial aviation to survive, we must travel less and fuel aircraft of the conventional variety with a cleaner high energy density liquid fuel, until such time as we are ready to fly in blended wing body airliners fuelled by liquid hydrogen.

Blended wing body aircraft

Using the perceptive modelling techniques, it appears that it may well be possible to achieve a 20% higher lift/drag ratio with a blended wing body shape than the best conventional shape now (Figure 10.3). In terms of emissions, we need that benefit. Such a shape has its problems. The pictures usually shown depict a cinema-style passenger cabin. The non-circular cross-sectional shape of this passenger cabin will result in high weight penalties. For long-range commercial flights, with zero emissions, we must

Figure 10.3 An early example of blended wing body design from Cranfield University.

fly using liquid hydrogen fuel, and this requires huge tanks. This leads us to place the liquid hydrogen fuel tanks in two banks, one on each side of a circular cabin. The fuel tanks can be placed within an unpressurized wing structure. We must avoid any hydrogen fuel leakage entering the cabin. This requirement fits well with tanks carried in unpressurized zones. The tanks must be segregated such that they can satisfy the requirement to balance the aircraft. The tanks are filled selectively to make sure that the centre of gravity for the loaded aircraft sits close to the centre of lift. This is vital for efficient flight to enable the aircraft to fly smoothly through adverse conditions. It means that the aggregate volume of fuel tankage is greater than the volume needed to carry the correct weight of liquid hydrogen fuel. Using the fuel to balance the aircraft is normal practice. For a given amount of energy, hydrogen weighs about one-third of the weight of jet fuel so the requirement for excess tank volume may be greater.

Liquid hydrogen needs to be carried in well-insulated tanks at a temperature of −253°C. For ground-based storage, and for the transport of liquid hydrogen, specialist companies know how to design the tanks with good insulation and safe systems to avoid leakage and minimize fuel boil-off. Moreover, their designs have systems to deal with these challenges. For commercial aircraft, we must replicate these safety features at a much-reduced weight. We can learn from our experience with space flight on how to design and make thermal insulators that are very light in weight. There is a need for innovation in the control of liquid to gas for hydrogen as a fuel.

There are other issues with the blended wing body shape. Without losing any of the aerodynamic benefits of the BWB shape, we must solve the problem of loading and unloading passengers and baggage. Unlike Maverick, the flight test model from Airbus (Figure 10.4), which has the centre line of the wing passing through the centre of the cabin, we need to lift the wing, such that the top of the wing skims the top of the pressure cabin. This must be done without losing volume for the fuel tankage. The lower skin of the wing rises steeply from the bottom of the passenger cabin, thereby providing space for doors through which passengers, baggage, and freight can pass efficiently. Viewed from the front, the aircraft will look more bird-like with the wing carried on top of the cabin.

Figure 10.4 Airbus' Maverick blended wing body aircraft.[3]

There are issues concerning stability and control, but these are solvable problems. Accurate modelling, flight tests, and wind tunnel tests are being used to optimize robust solutions.

Energy efficiency of commercial aviation

All the above address the goal of zero emissions with extreme urgency, but we must get there from where we are now. We must use the best carbon-neutral liquid fuels in the most fuel-efficient aircraft we already have. Early commercial flights using near carbon-neutral fuels have taken place. To do this across all fleets is a tough call, and the avoidance of a climate disaster is demanding better. New aircraft designs incur considerable emissions during manufacture, so maximum efforts must be made to keep these to the lowest possible level.

As always, new aircraft must be of improved aerodynamic design and as light as possible. Air traffic control and airport design for improved route optimization must be embraced to find the lowest fuel burn, and hence emissions.

We must live with the aircraft we have until the proper solutions are cleared and certified for commercial service. The tube and wing type of commercial aviation will be fuelled with a sustainable aviation fuel. The liquid fuel needs to be carbon neutral or as near to that as possible. Such fuels may be made on an industrial scale from waste, or from carbon dioxide direct air capture (as discussed in Chapter 5), then taken through processes that achieve the right properties, especially high enough energy density to match the aircraft performance on jet fuel. Airlines will need consistent standards across all airports to ensure safety.

If people must travel large distances, commercial aviation is the only realistic choice. Some journeys can be done by car, train, bus, or ship but it takes longer and needs more stops. It is not clear that surface transportation has lower emissions. Once we have zero-emission commercial aviation, surface transportation over large distances will be less attractive than now.

A petrol-powered car with four people on board achieves about 40 miles per litre per passenger, while an Airbus A350 flying nonstop with a full load achieves about half this. The plane takes a day. By car, if there were roads all the way, it would take three weeks. That takes a lot of stops, not to mention the coffee, cakes, and accommodation en route.

Air traffic control

Air traffic control is being improved to give more efficient flight paths. This can save up to 10% of fuel burn. The new digital control towers at airports give much more efficient flight paths. It also reduces pollution levels and noise at airports. Brilliant pioneering digital control of air traffic has been built at Cranfield Airport by Cranfield University (Figure 10.5). Their system is up and running. The capability of digital air traffic control is far superior to the old way of doing things. It can handle denser air traffic with a greater variety of aircraft with widely different needs. The picture shows the secure operational control room, which can be anywhere. It does not need to be in a lofty tower with a direct line of sight from the tower to the field of operation. A tall mast is placed at the optimum position and fitted with remotely controlled cameras, radar, and other perceptive instruments. The aircraft traffic controller has the means to see and analyse, in detail, everything that might cause a hazard. With the emergence of airborne taxi services covering distances of a few hundred miles, we will need the tighter air traffic control the digital control makes possible. If thought necessary, a controller could control a safe flight path all the way from take-off to

Figure 10.5 Cranfield Airport's digital air traffic control centre.

landing at the desired landing place. It is perhaps more likely that there will be many such digital flight control rooms with more precise handover procedures than the excellent procedures already in place now, as more journeys take to the air, many more air traffic controllers will be needed, but by going digital this becomes realistic, efficient, and affordable. Digital control means easier access for human air traffic controllers. The human exercising control needs absolute concentration and needs rest periods every couple of hours, making it important to have easy access from home. This is hard to achieve when the tower is remote from where the controllers live.

Already, the digital airport system developed at Cranfield University has been installed in several city airports, including London's City Airport.

Digital air traffic control systems can be adapted to suit farmers in controlling both arable and animal-based farming. One can see it being used by those building new infrastructure. An example might be a surprise discovery during the work requiring expert opinion from an expert many hours away. The site manager can home in on the item of concern to provide the expert with all the detail needed from which to make a sound decision on corrective action.

A quick word on trains

Many extoll the merits of electric trains using power from the grid delivered by overhead cables. This is often a good way of serving high-density routes, but it only meets our targets for avoiding a climate disaster when the electrical power is clean. This requires renewable energy sources to be supported by stored energy such as compressed hydrogen gas. Overhead power cables are not cheap, but when used frequently to carry high-density loads throughout much of every day, the cost is justified. However, if the cost per passenger mile becomes too high, it becomes better to use the hydrogen in fuel cells to provide the power. For some cases, the trains might be battery-powered provided there is a source of clean electricity for recharging the batteries. Batteries might suit an infrequent service, but the economics tend to make a weak business case. The track, stations, road, and river crossings involve costs that cannot be fully recovered from fares.

Studies have shown that hydrogen-powered trains might be sustainable for many low passenger number routes, especially if it improves citizen life quality and brings investment and work into remote areas. It is most likely

Figure 10.6 Alstom Coradia iLint, the world's first hydrogen train.

that we will see hydrogen-powered fuel cells on low-density routes. One of the compact hydrogen storage tank solutions, such as the Kubagen hydride tank, might offer the most practical solution for onboard and infrastructure supply lines, but we should not rule out liquid hydrogen as the solution where the routes are more heavily used. As well as being emission free, stations would be cleaner and quieter with hydrogen fuel cell trains. Many stations might generate their own hydrogen from renewables. Alstom is a leader in hydrogen-powered trains, well suited to important routes with light passenger traffic.[4]

As you can see, hydrogen-powered trains are running now, and these are building up valuable experience. As renewables increase, it is reasonable to contemplate hydrogen-powered trains on many routes served by diesel railcars. Trains operating at low hydrogen gas pressure (3 bar) sounds attractive but lost carrying capacity hurts costs per passenger mile. Can the lower fuel price offset that?

Summary

Commercial aviation is not a big offender in greenhouse terms, but it could rise in importance quickly as people want to travel far and wide, so it must be addressed, despite the difficulties. Trade between the Far East and the UK, Europe, and the United States is set to increase despite moves towards greater degrees of national self-reliance.

First, we must use technology to reduce the need for flying. Send the data, not the people. Second, we must move quickly to the use of a more sustainable liquid fuel. This must use the highest possible content of carbon-neutral fuels. We should use carbon capture to capture CO_2 for use in synthesized high energy density fuels. We must try to avoid using biofuels that need vast areas of fertile land to grow the feedstock for these fuels.

For small commercial aircraft flying short distances, we must start to grow strong businesses using battery electric and fuel cell plus battery hybrids.

We must also pursue zero emission blended wing long-range aircraft. There are many challenges, so we must intensify efforts to be ready for the increased urgency to make long-distance flying emission free. You can just imagine the birds at their lunch break saying, "We told you so!" We have benefitted from blended wing bodies since the first bird took to the air. The blended wing body arrangement might sit well with electric propulsion provided we make sure the fans can be fed with good air. There needs to be a clear path for air entrained by the jets from the fans. We do see pictures that are doomed to have a high base drag. I learnt how damaging that can be in terms of aircraft range from experience on fast military jet fights that used afterburners to boost the thrust required for supersonic flight. When using the afterburner boost, the engine requires the final nozzle to be wide open. For subsonic cruise, the nozzle must be reduced in area. Unless air can easily be entrained by the jet efflux, there will be a high base drag holding the aircraft back. This results in a higher fuel consumption rate and less range.

For long-range commercial aviation, we need to move quickly, but thoroughly, to hydrogen-powered blended wing body jets. The blended wing body will, most likely, be powered by two high-pressure ratio gas turbines with large propulsion fans sited above the wing to provide noise shielding. The gas turbines will be fuelled with liquid hydrogen.

Notes

1 ZeroAvia (2021), *Our Mission – Accelerate the World's Transition to Sustainable Aviation*. Available at https://www.zeroavia.com/ (Accessed 14 March 2021).

2 Vertical Aerospace (2021), *Vertical-Aerospace VA-X4*. Available at https://vertical-aerospace.com/va-x4/ (Accessed 11 March 2021).

3 Airbus (2021), *Airbus Maverick Press Kit Photos*. Available at https://www.airbus.com/search.image.html?q=maveric&lang=en&newsroom=true (Accessed 11 March 2021).

4 Alstom (2021), *Coradia iLint – The World's 1st Hydrogen Powered Train*. Available at https://www.alstom.com/solutions/rolling-stock/coradia-ilint-worlds-1st-hydrogen-powered-train (Accessed 11 March 2021).

Part 4

FOOD AND AGRICULTURE

Critical actions

- Eat less meat
- Accelerate the mass production and adoption of meat-like foods
- Farmers using technology to improve yield, quality, and price competitiveness
- Factory-grown food located close to communities

DOI: 10.4324/9781003193470-14

11

FOOD

DOI: 10.4324/9781003193470-15

During World War 2, German U-boats were sinking ships bringing food to the UK. Everyone understood the seriousness of that. Rationing was introduced and accepted by everyone as necessary. Some excellent studies were done to check on the effect of the reduced intake of favourite foods on the ability of people to work hard to support the war effort and to grow our own food locally. The results revealed that the rationed diet improved health and did not reduce the capacity of citizens to work hard.

Many families kept chickens. A large number kept pigs in their own gardens. My family, living in a semi-detached house in Coventry in the UK, kept 12 chickens and one cockerel. These supplemented our rations of eggs and meat. New chickens hatched from our eggs that were fertilized by our cockerel were raised at home. Our neighbour kept four pigs at the bottom of his garden. As children, my brother and I watched the feeding processes, the treatment of effluent on this neighbour's garden, right up to the slaughter by trained experts, and the initial butchering prior to curing the meat. Bedworth, a small town close by, became to be known as Bedworth the Bacon Town because so many families kept pigs at home in their gardens. It was not unusual to visit a house and find a side of pork already salted and hanging off the picture rail in the dining room. My dad was a skilled craftsman who was ineligible for military service, but his work contributed much to the war effort. Each day in the growing season, he cultivated our garden, which gave us most of the vegetables we needed for a good diet for two adults and two small boys. The garden covered an area of 20 × 5 metres, with an Anderson air raid shelter, half-buried, halfway down the garden. A good variety of vegetables were grown, many from homegrown seed saved from the year before. Each autumn, fallen leaves from nearby trees were gathered in old coal delivery sacks and then placed in the chicken run. The run would be one foot deep in leaves. The chickens would eat any insects. Their poo enriched the rotting leaves. In the spring, the rotted and compacted leaf mould, enriched by chicken poo, was used in trenches on the garden to improve the soil for healthy plant growth.

It may surprise many readers that food comes ahead of cars in terms of global climate damage with meat, especially red meat, heading the list of offenders. There are good reasons for believing that rearing animals for food may be much more damaging in climate change than all the world's cars. Methane, arising from the digestion of feed in the animal dissipates faster than CO_2, so the commonly used factor of 23 to 28 could rise to about 100 if we are looking at a ten-year period, or less, to the tipping point where a runaway process starts to gain momentum.

The "Meat Eaters Guide to Climate Change and Health"[1] rates each food item in terms of kg of CO_2 per kg of human food consumed. It is a fascinating guide that my readers may want to consult further. Lamb is seen to be the worst emitter of greenhouse gas, at almost 40 kg of CO_2 equivalent per kg of meat, with beef about 70% of that (27 kg), while chicken is just under 20% per kg (6.9 kg).

Research done in preparing this book has provided some surprises. Much of the focus of the discussion on climate is on CO_2 but methane climate damage is more severe than most people realize.[2] Methane dissipates far quicker than CO_2, but over a ten-year period, the climate damage is about 100 times more severe than CO_2. All animals emit methane from both ends. So serious is this problem that it provides a real incentive to find the processes that give the equivalence of real meat and dairy products without the random release of methane.

Cows

Much is written about methane from cows as they are serious contributors to climate change. As of 2019, there are 1.7 billion cows on Earth.[3] Each cow produces between 70 and 120 kg of methane per year, with an average of 100 kg/year. But methane is 23 times more damaging in climate damage than CO_2. This figure might well be too optimistic to represent the true damage because methane dissipates more quickly than CO_2. Taken over a 100-year period, methane is 35 times worse than CO_2, but this rises to 84 times over a 20-year period. Our horizon for critical action is less than 20 years, so we really should use a higher factor, for example, 100 times. Using the long-term factor of 23, cows contribute 3.91 billion tonnes of CO_2/year, while world cars produce 2.46 billion tonnes.[4] However, using what might be the more accurate figure for short-term climate damage for animal-derived food, the impact from cows rises to 17 billion tonnes. That is almost seven times worse than all the world's cars. It represents almost 50% of the world's annual CO_2 rate. If the conventional factor of 23 times CO_2 is used, cows still produce 11% of world CO_2. Even this figure is not sustainable.

Other meat-producing animals add their climate damage through emissions of methane, other gases, and poo. Many animals eat young seedling trees, thereby removing potential forests of trees that require atmospheric CO_2 to grow, thus removing CO_2 from the atmosphere.

Pigs

The (2019) world population of pigs is about 855 million.[5]

For pigs, the major concern is from poo that is commonly used as fertilizer on arable land. This causes methane to be formed as it rots away but it also leads to harmful run-off into our rivers causing big problems for wildlife and the steady flow of water through the river systems. Moreover, the methane that enters our rivers, lakes, and ponds releases this methane over time. Some recent studies reveal the release of methane from these water sources has a huge greenhouse penalty. I have not included this source in my calculations. Increasingly, pig farmers are pumping pig poo into lagoons that are covered and sealed, where the poo forms methane that is collected to be used to generate electricity. Pigs also burp methane, but at a lower rate than cows, and there are only half as many. Pigs add about 9.5% to the total for cows.

Sheep

Sheep also add to the methane damage problem. The (2019) world population of sheep is about 1.2 billion.[6] China has 13.2%, followed by India and Australia, each with about 6%. For each tonne of meat, the CO_2 equivalent output of CO_2 from sheep is about 45% worse than cows but a sheep only weighs 17% as much as a cow. Sheep might add 16% to that produced by cows. Sheep are particularly effective in destroying the young tree seedlings that could have grown into trees that require lots of CO_2 from the atmosphere to grow. I have not included this source of environmental damage.

Chickens

As of 2019, there are about 25 billion chickens.[7] They too add methane, but they create other nasties from a climate change point of view. Chickens increase methane/CO_2 by about 1% as much as cows.

Chickens reared in poultry houses make large amounts of ammonia. This is an issue for the health of the chickens and for people living or working close by. Ammonia reduces chicken weight-gain rates and harms egg production rates. Much of the chicken poo and pee is used on land as a fertilizer. This often leads to excess nitrogen in the soil. Through the nitrification-denitrification process, nitrous oxide is formed. Nitrous oxide is 310 times worse than CO_2 in terms of greenhouse gas worsening climate change. There are valuable studies taking place to address the minimization of this problem.

The total CO_2 equivalent problem created by the animals that world citizens breed for meat, eggs, milk, butter, and cheese amounts to about five billion tonnes, if we focus on the short-term damage potential, burping animals bred for food are up to twice as damaging to the climate as all the world's cars. The data used is subject to significant error margins, but at the worst extremes of error margins, the message remains clear the food humans eat is a serious contributor to adverse climate change and must be addressed. As with cars and aircraft, as more people move up the income scale, the climate penalty for the food these citizens want to eat adds to the challenge of avoiding a climate disaster.

Synthetic meat

We have noted the high climate damage globally from using animals to produce our meat, cheese, eggs, milk, and other dairy products There are scientific experts who understand how the animal makes what humans want to eat. Put simply, the animals use processes that can be replicated by engineers that can be used to build factories using processes that closely match the real thing. Once fully understood, it should be possible to do this at a price that ordinary workers and their families can afford. Some progress has been made but the unit price is still too high. There are several companies across the world, supported by serious investors and governments, who can see how to get synthetic meat products that most people want to eat for an acceptable price. Moreover, once consumption is at mass-market levels globally, it will be a major contributor to the efforts to significantly slow the decline towards a climate disaster.

In this fast-moving scene, there are a handful of companies worth watching (see below). This number grows rapidly as the potential market is huge. Avoiding a climate disaster requires the fastest possible progress to global mass markets, consistent with complete safety. The world has shown what can be done at speed with full safety paramount, such as in the case of vaccines for the COVID-19 virus.

- Beyond Meat (USA)
- Cell Agriculture Ltd (UK)
- Chr. Hansen (Denmark)
- Impossible (USA)
- Modern Meadow (USA)
- Mosa Meat (Netherlands)
- Upside Foods (USA) (Recently renamed from Memphis Meats)

Technically, there are two starting points from which a properly controlled engineering process can take place. Cells from an appropriate animal source are favoured by some. For the equivalence of meat from chickens, the starting point may be stem cells taken from the feathers of a chicken. Others prefer plant-based cells. Scientists at the University of Bath have successfully grown animal cells on blades of grass, which could be developed into cultured meat, such as chicken well suited to mass-market sales.[8]

In another example, researchers at the Israel Institute of Technology were experimenting with growing human muscle tissue for grafting when a PhD student requested permission to adapt their process to synthesize meat with a texture close to that of human muscle.[9] To achieve the right texture, a matrix structure out of soy protein was used, giving the cells something to grow on, which resulted in a muscle-like structure of the synthesized meat that was similar to the real thing.

There are cell-based food companies working on seafood. One such company called BlueNalu recently demonstrated its first commercial product, yellowtail amberjack. This synthesized fish can be cooked in many of the usual ways, with the added advantage that there are no fish bones or scales to worry about.[10]

While it is still early days, this pioneering work indicates that cultured meat and fish has a large mass-market potential. Cultured meat products have attracted serious funds from Bill Gates, Sir Richard Branson, and Kimbal Musk, the brother of Elon Musk, all of whom are investors in Upside Foods, of California in the United States. Upside Foods (previously known as Memphis Meats) has made progress in creating cell-based beef meatballs, chicken, and duck (Figure 11.1). It is likely that a preferred method of producing high-quality cultured meat will feature 3D printing and exercising the cell culture, as if it were animal muscle, to get the right taste and texture consumers will be happy with. Researchers are laying down thin layers of cell aggregates to build up a block of muscle that never lived. It is possible to control the placement of muscle and fat to attain the right taste and texture. Parallel work is taking place for cultured beef and other meats, including important work on fish. Barclays bank estimated, in December 2020, that the cultured meat market could be worth US$140 billion by the end of the decade.[11]

Laboratory costs are high, but with the huge appetite for meat in the United States, and most other countries around the world, many experts predict that volumes will reach a high level and lead to affordable prices.

Figure 11.1 Upside Foods – synthetic chicken.[12]

Research shows that the average citizen of the United States consumes 225 lb of beef per year. It has been shown how costly a ¼ lb beef patty is to grow in a cow. It takes 6.7 lb of grains, 600 gallons of water, and 75 square feet of land. Cultured meat will have 78–96% fewer emissions, 99% less land, and 82–96% less water. For many countries, an increased reliance on locally produced cultured meats and fish can displace a dependence upon imported natural equivalents. It is not surprising to see Singapore, with its excessively high dependency on imported food, setting out to be a world leader in the field of cultured meat and fish products. Other countries, including the UK, have an equal need to be among the world leaders. I cannot overstate the urgency to quicken the pace of getting to mass markets at the right price as the climate disaster shows its teeth through ever more frequent extreme weather events.

Engineering new food

It might seem strange taking a design engineering approach to creating new food to feed our global population, but in many ways, the principles are just the same as they are with any other complex system. We need to design for affordable success and safety within severe and changing constraints.

There appears to be a willingness to self-regulate, but in the interest of fairness, rationing might well be necessary.

Most UK farmers are aware of the need to reduce climate change damage from animals reared for food, but there are indications that the producers in these massive-scale businesses are still in denial. Cattle farmers in the United States still focus on getting costs down and driving up the consumption of steaks and meat patties. South America also appears to fit the denial model.

The previously mentioned figures take no account of where animal feed comes from. Much takes the form of soya beans. Large amounts are grown on land that was once forest but has been cleared by deliberate forest fires to make the land suitable for growing soya beans. While the growing beans need CO_2 from the atmosphere, which is helpful, the atmosphere may be overwhelmed by the loss of millions of trees that remove CO_2. This damage is further increased by the massive release of CO_2 by forest fires and the equipment used to farm and transport the matured beans to the animal feed stations. More than this, there are huge emissions of greenhouse gases involved in transporting animal feed over great distances. The net effect causes extra hurt to climate change.

There are many farmers worried by the prospect of consumers changing in huge numbers away from animal-derived food products. As noted earlier, those operating quality restaurants are already reporting a customer-driven trend towards smaller portions of red meat and an increased consumption of meat-like foods based on plants. Diners now are looking for the best dining experience rather than the biggest and most juicy steak. We can expect mass-market restaurateurs, competing to offer new and better dining experiences, to grow the market and restore market size and revenues. New technology will play a role in these developments. Farmers will likely adjust the way they sell their expertise. That is not likely to be an easy transition. The use of farmland may change to more intense crop farming and remote control of farming in urban areas. More land might be repurposed for leisure activities. As we will see later in Chapter 13, contrary to popular belief, we might find it better to live in more dispersed and greener communities with shared facilities, and a higher level of mutual support. This may be possible with less hurt to the environment than high-density high-rise urban living, as this creates other problems, including the need for food being brought in from afar, with high emission penalties, loss of taste, and a high level of food wastage.

We need a multi-pronged series of actions, including rationing, replacing animal production with synthetic foods, and alternative rewarding work for those employed in farming and horticultural industries. It would be good if we could avoid rationing but doing so places a huge reliance on people being adequately informed and willing to cut down on foods requiring animals. The UK imports much of its food from many countries, so the UK has the freedom to import less of the foods that are bad for the climate in three ways. Food from animals has a high greenhouse gas emission, as does much of the animal feed. Transportation of these foods and feeds adds emissions from transportation.

We must also develop the way animal-derived foods are presented, such that reduced and healthier diets become attractive to most consumers. This might be the way rationing is avoided. However, it does depend upon everyone doing what is right to an agreed set of standards.

Clearly, these are early days for synthesized meat products in what is likely to become a major industry across the world, with scope for many new jobs in this science-based new industry. It will be most disruptive for the conventional production of meat in animals bred for that purpose. However, countries with a high dependency on imported meat and fish will be able to become more self-reliant. This has a triple benefit in terms of greenhouse gases, as noted earlier. As COVID-19 becomes better controlled, the approaching climate disaster makes it clear that there can be no return to the old normal. With food raised in animals having such a powerful adverse effect on climate change, we must trim our natural meat intake and transition quickly towards cultured equivalents. As more and more humans move into "Level 4", animals bred for food will rise in impact on the slide into a climate disaster unless large scale corrective steps are taken promptly.

Summary

Collectively, animals bred for food are a major cause of climate damage. Action is urgently required. This may need the re-introduction of rationing and an accelerated drive towards engineered equivalent foods. There are signs of good progress, but until we get to a mass-market uptake, the price will remain too high, with a severe impact on climate change.

It is not just a question of price, however. We still have so much to do to help consumers realize that they too are part of this change. For too long meat substitutes have had a terrible reputation (mostly deserved) for lacking any redeemable quality other than they did not contain any animal products. That is yesterday's news. The new breed of meat substitutes is a world away from past experiences, and although younger consumers are open and often eager to try, we will need to do more to engage other consumers to take the first step. In my personal experience, once the first step is made, it's a much easier conversation.

Similarly, we will have to work hard with farmers and agriculture to ensure that they are not left behind, but instead, engaged and empowered to lead this change, which is what we will discuss in Chapter 12.

Notes

1 Environmental Working Group (2011), *Meat Eater's Guide to Climate Change + Health.* Available at https://www.ewg.org/meateatersguide/ (Accessed 8 March 2021).

2 European Commission (2021), *Methane Emissions.* Available at https://ec.europa.eu/energy/topics/oil-gas-and-coal/methane-emissions_en (Accessed 8 March 2021).

3 Food and Agriculture Organization of the United Nations (2021), *FAOSTAT (Data Repository).* Available at http://www.fao.org/faostat/en/#data/QA (Accessed 8 March 2021).

4 Our World in Data (2020), *Cars, Planes, Trains: Where Do CO2 Emissions from Transport Come From?* Available at https://ourworldindata.org/co2-emissions-from-transport (Accessed 10 March 2021).

5 Food and Agriculture Organization of the United Nations (2021), *FAOSTAT (Data Repository).* Available at http://www.fao.org/faostat/en/#data/QA (Accessed 8 March 2021).

6 Food and Agriculture Organization of the United Nations (2021), *FAOSTAT (Data Repository).* Available at http://www.fao.org/faostat/en/#data/QA (Accessed 8 March 2021).

7 Food and Agriculture Organization of the United Nations (2021), *FAOSTAT (Data Repository).* Available at http://www.fao.org/faostat/en/#data/QA (Accessed 8 March 2021).

8 BBC News (2019), *Artificial Meat: UK Scientists Growing "Bacon" in Labs.* Available at https://www.bbc.co.uk/news/science-environment-47611026 (Accessed 8 March 2021).

9 Engineering and Technology (2020), *Israeli Scientists Create Fake Beef Scaffold*. Available at https://eandt.theiet.org/content/articles/2020/03/israeli-scientists-create-fake-beef-scaffold/ (Accessed 8 March 2021).

10 BlueNalu (2019), *BlueNalu Highlights Versatility of Its Cell-Based Yellowtail Product in Premier Culinary Demonstration*. Available at https://www.blue-nalu.com/pr-121719 (Accessed 8 March 2020).

11 Barclays Investment Bank (2019), *Carving up the Alternative Meat Market*. Available at https://www.investmentbank.barclays.com/our-insights/carving-up-the-alternative-meat-market.html (Accessed 11 March 2021).

12 Upside Foods (2021), Real. Delicious. Meat. Available at https://www.upsidefoods.com (Accessed 08 July 2021).

12

FARMING AND AGRICULTURE

DOI: 10.4324/9781003193470-16

There were few kids for me to play with in my infant and junior years, so I had a lot of time to make models, garden, and think. I liked sums. I realized that it took more land to feed a person than to house that person. I knew what we grew in our garden was not sufficient and we were surrounded by fields of wheat and barley. Close by was a flour mill driven by a water wheel. It made me realize that as the population in cities grew, food would have to travel further. How do you keep it fresh and safe on its travels? It was ok for my family. We had a productive garden and chickens to provide our eggs. Our milk was delivered by the local dairy farmer in churns transported by horse and cart. The six cows that provided the milk for our few local families grazed on several big fields. Was there a way of getting more food from each area of land? In later years, I realized how our diet depended on a variety of foods. Each variety needed different growing conditions, so waiting was part of the process used by nature to grow our crops and provide feed for our animals kept for food. Later still, I felt that nature had given the tools for people to study what plants want at each stage from seed to germination, and onwards to maturity, harvest, and packing for travel to the consumer's table. This shouted factory! In my neighbour's greenhouse, two layers were used for many crops. Then I realized cleverer folk had already started to use multi-layers as well as just-in-time accelerated growing techniques. This meant that with a much smaller footprint and accelerated growing, year-round fresh foods could be closer to where people lived, matching the predicted demand. Without new technology, the costs would be too high. Efficient LED lights produce only light of the right wavelength and recycled materials hold the plants while they are fed and watered to match their demand.

Many of our vegetables, fruits, and salads are grown on farms or in huge greenhouses. Many are imported, often involving many miles of road transport. All these methods have at least a partial reliance on Mother Nature, especially in the form of light, temperature, microbial activity in the soil, and weather. Farm grown food needs fertilizer to increase yields and pesticides to reduce crop reduction through pests. Pesticides have an adverse effect on pollinating insects. Fertilizers often result in run off that pollutes streams and rivers. As more and more people move to live and work in big cities, this urbanization leads to food deserts. In turn, this means food must travel further, usually by land transportation to get from the farm to local shops/supermarkets. This transportation produces large amounts of CO_2 and other pollutants such as fine particulates that harm the lungs. Moreover, quality and taste are lost in transit. Food waste increases, especially when travel time is long. In the drive to raise synthetic meat quality while lowering the price to mass-market levels, there is scope for much innovation.

Food plant-growth factories

We need to grow our plant-derived foods closer to where the food is consumed. We also need to harvest the mature plants at a time that matches demand. It takes thorough work by experts to find out exactly what each variety of plant needs, and when it is needed during the process from seed, through germination, growth to maturity, harvest, and packaging for delivery at the right time and place. Much good work has already been done in many countries. Many investors controlling big sums of finance are showing interest. Sadly, some are losing money by investing in companies that do not have the right skills, and the well-researched data, to deliver a satisfactory return. In every walk of life, the real prized result takes hard work and lots of it.

Some farmers are losing productive land through flooding. This is happening more frequently in recent years. This suggests the need for food plant growth on more scarce high ground, leading to a need for more intense factory farming. We could also consider food-growing factories that float, such that food production can continue during flooding and the time lost for cleaning up after the floods have receded. With seasonal flooding predicted to become more frequent, with the number of areas also increasing, it is time to evaluate floating homes and factory farms. There is also the prospect that renewable energy powered floating food factories will suit the need of communities living on water or close to the sea or lakeside locations. There might be a business case for old ships to become solar or wind-powered food farms once their sea-going life becomes uneconomic.

Renewable electricity for food factories does require energy storage to ensure uninterrupted growth to the correct schedule. As noted in Chapter 7, it may be possible to achieve more electrical power from large numbers of small safe nuclear reactors that address the fear factor that has held back the nuclear industry for many decades and do this at a low cost per kWh.

Food-growing factories offer a real prospect of increasing yields, quality, and taste, at the right time and place in purpose-built factories, using a combination of multi-layering to gain growing area/unit footprint on land, just-in-time delivery of exactly what the plant needs, and robotics to do the verification of the process plus the lifting, harvesting, and packaging.

Case study: Jones Food Company[1]

The Jones Food Company, based in Scunthorpe, England, is a well-respected front runner in food factories. They have received support from the Ocado Group, the UK-based grocery retailer/tech company that develops software, robotics, and automation systems for online retailers around the world. They have focused on salad crops. They also are growing plants used in the pharmaceutical and cosmetic industries. They use high-intensity light, at the optimum frequency, for the optimum number of hours per day. The Jones Food Company has one of the largest, if not the largest, operational food growth factory in Europe

The factory is owned and operated by the Jones Food Company and has 17 layers. Plants are fed hydroponically while growing on a bed of Rockwool insulation fibre under intense pink light in a sterile atmosphere rich in CO_2. There are many variants of vertical farming, but they all share a common theme of raising productivity per unit footprint, raising food quality, and matching production to the time and place of consumption, at the right price with reduced environmental damage.

With plant-growing food factories, we can expect to avoid pesticides completely. There will be a 90% reduction in the use of water since the water is recycled within the closed and sealed factory. By directly feeding plants with CO_2 harvested from the atmosphere, produce grows more quickly, reaches a higher weight at harvest time, and less fertilizer is used. The growing medium can be a wool-like mattress, derived from shredded and recycled plastic bottles. Feed for the plants can be delivered in vapour form directly to the roots projecting out below the medium holding the plants. In other examples, the plant food and water are delivered in liquid form to a tightly controlled programme.

Most important is the control of lighting both in terms of wavelength and duration.

Case study: John Innes Centre[2]

Researchers at the John Innes Centre have studied just-in-time growing of specific food crops and found that for the plants studied they could get six crops/year in place of the normal one per year. Other researchers elsewhere across the globe have results that reinforce confidence in the work at the John Innes Centre. The John Innes Centre research includes important plants like barley. Some plants are relatively easy to grow in these specialized factories. Others have more complicated requirements and for the near future will still be grown by conventional methods. Salad crops and soft fruits are among the front runners while potatoes might need more time.

Case study: Intelligent Growth Solutions[3]

Intelligent Growth Solutions (IGS), based in Edinburgh, Scotland, has also made some outstanding advances. In its food factories, multiple layers are used, together with just-in-time feeding and watering (Figure 12.1). Most importantly, they are using LED lighting with the right wavelengths of light, best suited for the optimum growth of the plants for a larger fraction of the day than nature can provide. IGS's excellent results are attracting serious attention from supermarkets. The company has tightly controlled patents covering much of its expertise. Their system has a high level of automation to ensure consistent high quality and high productivity. Year-round

Figure 12.1 Intelligent Growth Solutions CTO and founder Dave Scott standing next to a 9 m tall growth tower.

deliveries of strawberries and other fine foods are achievable, wherever, and whenever, they are wanted by consumers.

We need to recognize that many other countries, such as the United States, the Netherlands, Spain, Italy, and China, are already active in these technologies and the engineering follow-through needed to meet mass markets locally on a global scale. The nuclear power station explosion at Chernobyl in 1986, which caused large areas of contaminated land, may have prompted much innovative action to retain adequate food supplies should there ever be a repeat of such a nasty accident. Plagues of locusts is another hazard growing in importance which we'll briefly discuss later. Food-growing factories, with their tightly controlled growing environment, are completely robust against this kind of hazard.

Another by-product of the indoor vertical farming approach is that it can help to drive up national food security by enabling more effective use of land to gain national self-sufficiency.

It is interesting to realize that there are other big benefits from researched and engineered food-growing factories.

In terms of productivity, factory growing of food can be at least 24 times what we achieve now (six crops/year times four layers). With 17 layers at a food-growing factory like the Jones Food Company, we might achieve two orders of magnitude in food productivity per unit of land.

The new ways of farming will find their natural place in the market. It is unlikely that food grown in factories will fully replace the improving ways of conventional farming, but the combination of getting the highest quality food at the right place, at the right time, and at the right price will have a big impact, especially in urban areas, remote areas, and locations in distress.

Plant-growth factories in urban areas

As old industries die out, especially within urban areas, old factory and retail sites become available. While these can be pushed over to make room for increased living space in urban areas, climate change alleviation suggests that it might be better to repurpose these old factory buildings to be used to grow the food needed locally. Equally, the technology used by most of us to connect with the rest of the world can reduce much of the daily commuting by office workers, thus freeing up more office space in dense urban districts, which could also be redirected to food production locally, across the globe.

"Growing Underground" in London (www.growing-undergound.com) provides an excellent example of this where a team of entrepreneurs have turned an old unused underground station, last used as an air raid shelter for Londoners during the World War 2 bombing raids, into a fully operational underground urban farm (Figure 12.2).

These ideas must be thought through very thoroughly. It takes much more land to feed a person than to house that person. Growing locally makes real sense, and climate change alleviation necessitates urgency for the global uptake of this more productive form of farming. It has lower emissions and the potential to deliver great freshness, great taste, at the right price for many of our common foods.

In Berlin, Germany, Infarm – a start-up company – is co-locating vertical farms with supermarkets and shopping malls (www.infarm.com). Also, in Berlin, there are in-store micro-farms in Metro stores where customers can buy growing plants ripe for harvesting in order to give freshness and fullness of taste. Glazed cabinets with access doors are

Figure 12.2 Growing Underground – growing salad under the streets of London in a disused underground station.

placed fresh each day, as required, alongside the displays of fresh fruit and vegetables.

This trend for growing locally, globally, is gaining momentum in the United States. Once we get to high throughputs, the economics should start to look attractive to investors and customers. The extra taste and consistently good quality, timed to match customer demand, will accelerate the uptake. There are indications that leading retailers are starting to see the potential and the fast pace of increasing customer demand. Social media will likely play a role in driving this fast-paced change to embrace the transition to much more food factory farming. Factory-grown food produce can significantly reduce the food waste that occurs at the supermarket, and later, at home. Fish farming might also be embraced in this movement. The fish can also be bred to achieve quality and good taste, while their poo can be used to fertilize growing plants in factories.

With all the light coming from electrical power, and the need to power robots and other machinery, there are real concerns about cost per unit output. Some argue that only the wealthy nations can afford these costs, thus making it too costly for countries needing such farms the most. However, countries in the sunbelt have the potential for extremely cheap solar electricity. We just need to take a total systems approach, combined with a large increase in the movement towards indoor vertical farming.

Factory and vertical farming in urban areas may well be set to increase in scale in most urban areas. Technology makes it possible for rural farmers to control food-growing factories, in urban areas, in a way that protects their income.

Onshore wind turbines require huge areas of land to get a worthwhile output. Could we not drive up income on that land by co-locating food plant-growing factories? This would be doubly beneficial where wind turbines and solar farms are close to where people live and work. Great thoroughness in design would be needed to avoid vibration and noise increases from the turbines that might arise from disturbed air passing through the wind turbines. As with smart cities, we might place the vertical farm buildings in an array that increases the harvesting of wind power.

The discovery of new materials is raising the prospect of low cost, solar sheets that are flexible and can be laid on any suitable surface. These discoveries lead to inventions that enhance the productivity of land in terms of both food production and renewable power generation.

As noted earlier, farmers might also want to run urban food factories remotely from their rural abode, noting that robots and automation might well take the bulk of the effort. There might well be mutual benefits for operating in this way. Remotely controlled cameras, drones, and scientific instruments can provide good data from which to get consistently high quality and cost-effective results that satisfy the farmer, customer, and investor. Some will say factory farming reduces jobs, but in reality, it means nations can grow more and import less.

Ground-based robots can undertake routine tasks like selective weed killing without using nasty chemicals. They can detect early signs of problems and deliver appropriate remedial action, all of which enables crops to attract the highest yields and prices.

Plagues of locusts

Several countries in Africa face famine arising from huge plagues of locusts that consume all their food crops. There is an urgent need to restore food supplies to the millions of people that have lost their food crops. Short-term relief requires massive airlift operations that are not sustainable. It is technically achievable to quickly build all the food-growing factories needed to secure ongoing food supplies the locusts cannot reach. They do need electrical power, and while the countries involved are all in the sunbelt, the factories do need heat and power for more than the hours of daylight, so energy storage is vital. Once many standard units are built, the growing process can be controlled remotely. This might require the digital airport control tower technology, already discussed as demonstrated at Cranfield University Airport. This would enable a farmer, living thousands of miles away, to control everything with the help of a few local people, aided by automation, robots, and big data. The concept needs some development, but the experience would be gathered very quickly. Moreover, it paves the way for farmers with experience, living in remote farmland, to control the output of grown food in many urban farm factories, with only occasional travel to the urban factory site.

People with the knowledge to do this are with us now. There is cash presently making negative returns. We need the right people to come together to make it happen. It has huge revenue-generating potential while strongly adding to climate change mitigation. It is a large jump for people to grasp but we have embraced other big changes in the past. The pace of change is quickening. Change is the new norm. Getting left behind will not be a happy experience!

Notes

1 The Jones Food Company (2021), *Vertical Farming for All*. Available at https://www.jonesfoodcompany.co.uk/ (Accessed 8 March 2021).

2 John Innes Centrse (2018), *Speed Breeding Technique Sows Seeds of New Green Revolution*. Available at https://www.jic.ac.uk/press-release/speed-breeding-technique-sows-seeds-of-new-green-revolution/ (Accessed 8 March 2021).

3 Intelligent Growth Solutions (2021), *Ideal Conditions for Life*. Available at https://www.intelligentgrowthsolutions.com/ (Accessed 8 March 2021).

Part 5

HOMES AND HOUSING

Critical actions

•   Accelerate the move towards better, affordable, carbon-neutral housing

DOI: 10.4324/9781003193470-17

13

URBANIZATION VS DISPERSED COMMUNITIES

DOI: 10.4324/9781003193470-18

As I grew towards maturity, it was clear that a good education gave better job prospects, but those jobs were in more distant locations, and families started to become less able to provide all the support from within the family. As a result, more care for infants, the ill, and the aged had to come from the state or paid for out of savings. This led to the need for mutually supportive communities, with easy access to all services. From the early 1960s, I travelled extensively around the United States to visit aircraft designers and manufacturers, often staying in motels made of wood but graced with facilities for relaxation and basic services. Many were set in areas surrounded by trees. Fire safety was easy as most were low rise with multiple easy escape routes, backed by clear instructions on the inside of the door. I was often entertained in the homes of the design engineers I had come to see, and many lived in large cosy sheds with a brick veneer to give the favoured look. Many of these homes were built from factory-made modules and erected on a concrete base and were up and running in a few weeks, complete with all fittings and furnishing.

Why are we building our houses one brick at a time in all weathers? Most other industries have moved on to improve function and reduce cost. As populations rise, the rate of building new houses and communities must increase substantially to recover from the shortfall and match the rising demand for better houses at prices supportable out of earned income. We have the technology, and most nations need the jobs.

If we can mass produce cars with increased choice, we can do the same for houses. We can print them with a 3D printer, or press and weld them from clean steel, and we can get the price right, provided people choose the new technology offerings and help improve them further to meet yet higher standards.

Urban living, in high-rise buildings, is neither good for climate change nor the health of residents. Dispersed living among increased greenery is now possible, with positive benefits in terms of climate change and personal health supported out of earned income. This can be done in ways that retain the natural beauty of the countryside. For those continuing to live in high-rise apartment blocks, there is a need to green them with vertical gardens of trees and plants like those in Singapore using an approach called "biophilic design". A biophilic building attempts to replace walls, windows, and columns with leaves, bark, birds, and insects. WOHA (www.woha.net), an innovative architecture practice in Singapore, shows some impressive examples of what can be done when designing buildings that establish new relationships between nature and the manufactured world. The greenery absorbs much of the pollution and helps moderate temperatures. Bare buildings release too much heat and add to global warming. They also increase the need for power-hungry air-conditioning. The "greening" of the built environment needs a rigorous maintenance programme as nature cannot always manage itself, especially in a human-made environment. But experience in places like Singapore shows it is possible and easily manageable.

In Southeast Asia, there are massive programmes focused on smart solutions for smart cities, with many new smart cities in China and India coming online. More than this, there are active plans for several pilot smart cities in each of the countries in Southeast Asia. Great attention is given to harvesting renewables through solar panels integrated into the structure of buildings, including high-rise residential or service company blocks. The high-rise buildings are carefully placed to maximize the harvesting of wind power, often using combinations of helical twisted vertical turbines.

All transport is either human powered or electrical derived from local renewables including energy storage. All services are managed for optimum performance, with tight controls on emissions, waste management, efficient use of power, and the control of water quality and supply. Jobs and the most mobile people will migrate to these new cities. Modern-day communications will keep those who have migrated in touch with those who have stayed at home.

In densely industrialized nations there remains scope for correcting these shortcomings and to provide indoor vertical farms that can match the local demand for fresh food. The demise of city-centre shopping brought about by the increasing dominance of online shopping provides realistic opportunities to embrace smart solutions that make cities great places to live, work, and play. Even smart cities make greenhouse gases. While some can be collected and used, the remainder needs to be absorbed by trees that deliver the oxygen cities need for healthy lives.

New homes are much needed that are complete with features better suited to the new ways of living and are required at prices that are affordable from earnings. Such homes need to be carbon neutral or better. We know how to do it but the uptake by developers and house buyers is too low, wasting the emergence of a substantial stream of technology-assisted employment. We need to re-think the balance of merits between high-density urban living versus dispersed living in a greener environment that retains all services, with most delivered in a new digitally enabled way. Existing high-rise apartment blocks need more greenery on and around the building to give better control of heat from the sun and heat leakage in winter. Moreover, trees and shrubs absorb CO_2 and gather fine particulates on their leaves that get washed off in the rain to form a less harmful sludge. Singapore uses this approach. All these trees and shrubs are well-managed to prevent damage or harm to citizens.

There are few signs that the established developers will follow this path in the mass housing market, so the new entrants to the building of new community developments may have to take the lead. We know how, but we need a change in mindset to bring it to reality at a large enough scale. There is a need to educate everyone that a carbon-neutral home need cost no more than a traditional home. Indeed, an environmentally friendly house can be better and cheaper than traditional designs. It is likely that when we have better housing, enabled by technology, the big players will want to be part of the change, adding their hard-won experience to meet the demand for affordable housing in ways that grow their businesses. There remain big challenges ahead, such as overcoming resistance to change from what people are used to.

The leadership may come from Saudi Arabia, where they have the NEOM project. NEOM is a planned cross-border city in the Tabuk province of northwest Saudi Arabia. This project is scheduled to start in 2021. The concept involves a radical fresh start to city design. The city takes the form of a single straight line 170 km long, served by a two-way ultra-high-speed rail link. It is designed so that no one needs to walk more than five minutes from home to work, or to reach services. The city will be powered by a green energy business. It has a dedicated artificial moon for lighting. It has provisions for making rain from atmospheric moisture overhead. Many human activities are robot-assisted. The founder of this imaginative scheme is Mohammad bin Salman. The region covers about 10,000 square miles. This concept for city living will stimulate similar schemes elsewhere around the world. In some ways, the basic concept is not new but taken as a package, it brings together many advances that work well together, from which citizens will see big gains, with reduced emissions. Many of the world's citizens will see this new city. It is well placed geographically.

Smart or not, is more urbanization the right way forward?

Following the Industrial Revolution, the number of workers commuting from home to work, for up to four hours per day, has increased enormously. Many millions do it, in every industrialized country and many agriculturally based communities. While many commuters use public transport powered electrically, the electricity used is not yet clean and many private cars are used to get from home to the railway station, often using large, thirsty 4×4s.

Many of their homes are old and thermally inefficient. There is also a shortage of good new homes that are more efficient, and better suited to modern living. Sorting this out is urgent, as many of the benefits from working from home, such as more time to look after the affairs of family and friends, will be lost in emissions of greenhouses gases that overwhelm the gains from reduced travel emissions. Millions of affordable new homes are needed to service the shortfall. More than this, too many homes are simply not fit for purpose, especially when working from home becomes more prevalent. We are wasting opportunities to embody renewable energy harvesting materials and systems into the building structure and community settings for new and rejuvenated old buildings such as disused places of work like city-centre retail shops.

There have been many years when the build rate failed to match demand, leading to high prices, and possibly robbing us of investment funds to increase productivity. The unproductive travelling time devoted to commuting leads to extra demand for state-provided services, that in earlier times would have been provided by family and friends. New communities in dispersed locations are required to house mutually supportive families, in a green setting, in harmony with the local landscape, and supported by all the infrastructure needed for efficient living in a clean environment. Some of the present infrastructure could be replaced by technology that enables personal travel to be replaced. We are used to online shopping, and medical advisory services, and more examples are emerging. Lockdown in connection with moves to control the coronavirus pandemic has highlighted the merits of working from home, but this has thrown light on many new problems that need to be fixed to provide healthy and happy lifestyles.

Designing better communities #1: Mutually supportive communities

Earlier we looked at smart solutions for smart cities. A parallel approach can be taken to greener, mutually supportive communities more dispersed into areas not previously seen to be viable. Now with new technology, people can work from anywhere. Many will choose a more rural setting. The challenge is one of working in harmony with nature rather than encroaching on nature. As many families no longer live close together, some of the self-help common in times long since gone must be provided by the state or services paid for out of pensions or savings.

Maturing children reaching adulthood relocate to find work for which they have trained, and where they can live at a cost supported by earned income. Children and the old need support that once came from within the family or mutual support from friends and neighbours.

Technology makes it possible to work internationally from a suitable place of work close to home. The coronavirus has forced millions to stay at home, with those still able to do their work working from home. Although I have done this to a minor extent for 30 years, I am not keen on working exclusively from home, as life inevitably gets in the way. Instead, it may be better to have dedicated places of work that are fit for effective work and provide the support services needed, including provision for socializing, sport, and entertainment.

These dedicated places for work need to be secure and properly supported with services close to home, ideally without the use of powered personal transport. Walking and cycling would be possible for many, while for those less able, clean-powered personal transport would be used. In Shanghai, many people, including many old people, ride electric bikes, with batteries charged at home.

Students, too young to earn a driving licence, swell the number of riders on electric bikes, trikes, and electric scooters. Initially, electric scooters were toys for children, but now there are many places where the "suited and booted" salary men and women use heavier duty electric scooters for some of their commutes between home and work.

Increasingly, the urban electric cars that we discussed in Chapter 9, with some in the quadricycle or motorcycle classes where no licence is required, are encroaching into the moped and scooter market. These offer weather protection. Speeds are low and the range is short consistent with urban runabout duties.

Despite my personal preference for working close to home, rather than working at home, for many millions working from home is the best choice, but it will take some time to learn how best to do this. It is an enormous change and one loaded with challenges to make sure the gains are not offset by health problems, especially mental health issues.

The change affects every member of the household. It is important to understand the issues so that problems can be avoided, and the full potential gains can be enjoyed. We have a great need for more homes, and if a high proportion of workers are to work from home there is scope to redesign homes to make them better suited to distance working.

In a typical family of mum, dad, and one child working remotely with a university and another online with school leads to the need for extra compartmentation, and extra heating/air-conditioning, not to mention industrial-strength Wi-Fi! All four will often be working at the same time. It is important to ensure heat leakage is kept low to avoid high emissions and costs. Locally harvested renewable energy converted into more usable heat and power help solve the problem.

Once such systems are rolled out in large numbers of standardized modules, they should become more affordable and supportable out of earned income. On a bespoke basis, prices would be too high. We must use high-volume standardized modules that can be joined in many ways that give variety without taking costs too high.

As well as being affordable, the new homes need to be made using methods of construction that are cleaner than current labour-intensive methods using materials produced with high emissions. Such new homes need low environmental impact.

Affordable housing for mutually supportive communities

The enormous shortage of affordable homes is starting to be addressed but many of the conventional construction processes are inadequately supported by skilled workers, such that there is no chance of meeting the need using traditional building practices. While these best efforts should continue, they must be supplemented by other methods of construction. These must include the use of timber from sustainable sources, built in a modular way in dedicated factories. Legal and General has entered the house-building business and already has the largest assembly hall in Europe for building modules for new homes.[1] Initially, the new method is not profitable but as volumes increase and methods of manufacture improve, costs per unit will fall substantially for homes with higher quality, better functionality, and with greater inter-family mutual support resulting from a resort-style of living while paying for it out of earned income through a mortgage or

Figure 13.1 Legal and General's house assembly hall.

rental arrangements. The new factory in Yorkshire is led by Rosie Toogood, an experienced manufacturing manager attracted away from a senior post within the Rolls-Royce Aero Engine business. Other highly rated manufacturing engineers, many from aerospace, are being recruited to improve build process effectiveness and efficiency while lowering costs as volumes rise. For me, this is another good signal that the skills of the design engineer are crucial to our collective success.

Until recently, many countries were ahead of the UK in the use of timber in the modular build of new homes. Germany, the Netherlands, Austria, and Scandinavian countries figure strongly, but there are others. These efforts sit well with the need to grow more trees but to scale to bigger volumes, they may not be fast enough to contribute strongly to climate change mitigation, but it is hugely important for its direct benefit in terms of quality homes in mutually supportive communities, funded out of earnings, and for those no longer facing the time-consuming daily commute with obvious reduced costs.

Cross-laminated timber (CLT) features strongly in modular homes built from sustainable timber resources. The trees need CO_2 taken from the atmosphere to grow. Even when processing is included, it is a good material compared with steel and concrete. CLT has a high strength/weight ratio. Indeed, serious consideration is being given to constructing a very tall building in central London. There are examples of high-rise blocks of flats built from CLT panels. CLT is particularly well suited to resort-style dispersed living where the buildings might be only two stories high to live closer to natural surroundings with trees, plants, gardens, and water. Many of the early US motels found in the post-war 1950s used this type of construction. It is especially safe from the fire risk point of view. Multiple escape routes are easily provided in low-rise, resort-style buildings. It is also conducive to good neighbourly interaction. It provides a calming ambience and encourages walking, cycling, and other good forms of exercise. More trees, and greenery, help to reduce the hurt to natural landscape beauty while reducing the concentration of CO_2 in the atmosphere. CLT can be faced with solar panels at the factory where the CLT modules are made.

It's worth noting that fire regulations arising from the serious loss of lives at Grenfell Tower in the UK and other similar fire events around the world have led to restrictions on the use of CLT in wall structures used to carry loads. This has undermined the market for such buildings, however, there are potential solutions, but the economic case has been impacted.

Three-dimensional printed homes from concrete/resin

Most people know concrete production requires lots of energy and conventional methods give rise to unreasonable levels of CO_2 emissions, but there are innovations that drastically reduce CO_2 to close to zero. Heat can come from renewables in the form of hydrogen and processes that use recycled CO_2. More than this, we now have 3D printing to give us much more efficient use of concrete. Stronger and lighter structures can be made using 3D printing. More efficient enclosures are possible, and the build time is down from months to about one day.

Three-dimensional printing is gaining ground rapidly from its early use by major companies making parts for aerospace and automotive industries. We have already seen 3D printed applications in meat production from animal and plant-based cells where there is a need to get the structure right. A structure of meat that is closer to animal-raised muscle tissue is required.

Work has already started using a fast-setting form of concrete to make modules of buildings, using far less material than that used in bricks, mortar, steel, and concrete in conventional builds. Computer-controlled printing machines place a continuous flow of material that quickly builds up modules or even complete houses on site, one thin layer at a time. The gain comes from using less material, and from increased air spaces designed to reduce heat losses. These gains are further increased when the concrete is made using heat from renewable power, and power for mixing and the printer comes from renewable sources. Provision is made to receive appliances with their need for electrical and plumbing services, complete with access panels needed to deal with any failures that can happen over a long design life for the building. Home power generation using solar panels and wind turbines can be an integral part of the finished house. Heat pumps and energy storage add strongly to self-sufficiency and minimize demand from the national grid.

In southern Mexico, a US-based charity called New Story (www.newstorycharity.org) set up a new initiative to 3D print very affordable homes in Tabasco, one of the poorest areas of Mexico.[2]

New Story was formed in 2014 to pioneer new solutions for homelessness globally. It has built more than 2700 homes by conventional methods in Mexico, Bolivia, El Salvador, and Haiti.

At the end of 2019 in the outskirts of a town in Tabasco state, they completed the first homes in the world's first-ever 3D printed neighbourhood. By European standards, the houses are small, with a footprint of only 500 square feet. Each house took 24 hours to build. Each house has two bedrooms, a living room, a kitchen, a bathroom, and a porch. The floors and walls are printed in a cement-like material known as Lavacrete. Residents pay a mortgage of US$21 per month over seven years. A median salary is US$76.50 per month.

Figure 13.2 shows the home. These small houses were driven by the need to home millions of homeless people, across the world, hence the small footprint. However, there is nothing to prevent housebuilders from making much bigger homes, perhaps taking a modular approach. Renewable power harvesting combined with energy storage can be part of each house or placed between houses. The homes are also earthquake resistant. The 3D printer called Vulcan 11 is 33 feet long and is robust enough to work in disaster zones.

A 500-square foot home is too small for many communities, but a two-story building with that footprint gives 1000 square feet of usable space, and that equates to the size of many much-valued average homes in the UK or many other countries around the world.

Figure 13.2 New Story's 3D printed home in Tabasco, MX.[3]

The message is clear. The way is open for building larger homes much faster than can be achieved by conventional build techniques, and the new homes can be efficient with excellent facilities of high quality while creating much-reduced greenhouse gases. We can expect to see developments along these lines in population centres with a high need for mutually supportive communities living well in affordable homes. Meeting the pent-up demand for affordable new homes using high technology-based processes will bring many new jobs despite the much-increased use of automation, including robots. The new modular homes can provide the right spaces for several people working at the same time with multiple remote connections; 3D printing gives greater freedom to optimize house design. Exotic shapes are possible for those with a forward-looking flair.

This example again shows the importance of innovations coming from anywhere on Earth. If it is the best, let us access that and adapt it to specific needs. It also needs the correct material, but that can be found. Find the best and make it better. Will the well-renowned building companies run with this? Perhaps a university-based start-up or more charities might pose realistic challenges in the mass market. My expectation is that the combination of climate change and recovery from COVID-19 will encourage the building industry to embrace the faster house building processes, as affordability and the need for a more streamlined delivery of services become vital. We must avoid the return to our wasteful way of living.

There are many references in the public domain, accessible via the internet, which reveal the scale of action across the world in projects that provide homes for the homeless, right up to prestigious homes for the rich.

Three-dimensional printing has the potential to build structures previously thought impossible. For example, taller wind turbine towers offshore would permit larger and more efficient wind turbines.

Homes built from metal

There is a wealth of experience in temporary homes and offices for which the cost and speed of delivery must be kept low, while the functionality must be of the highest order. Many are made of thin gauge steel and fit within the envelope of a standard full-size container, such that they can be transported by road without an escort. Modules within this size are ganged together to match specific needs for space and services. I have excellent experience of a large integrated assembly of offices in the car park of beautiful downtown Burbank Airport, housing big teams of designers pulled together from many sites to design a powerplant integration with a brand-new Lockheed Tristar wide-body jet. These buildings were comfortable and fitted out to a standard higher than was normal in older buildings. Since this experience in 1968, many improvements have been made. Container-sized modules can be delivered to the site as a flat pack that can be readied for occupation within a little over one day. Advanced design and manufacturing techniques provide the means for meeting most demands in a fast timescale. They have the potential to meets the needs of the homeless. They also can be made to suit the high-end market. Most importantly of all, they can achieve low cost, even when designed for a life of many decades. The stained image of prefabs and portacabins has no place among this class of housing. The steel used must come from clean sources using heat from renewables, usually green hydrogen.

Aerospace demands lightweight, with every ounce of material working hard and close to its safe limit. It might make good sense to use skills developed for aerospace, automotive, and white goods industries to use thin sheet steel to make ideal, functionally effective, well-insulated modular, and affordable homes. These homes, especially as modules would be light in weight, are well suited for transport from the factory to the site selected for the new home. The modules might be container size but made to high standards of functionality and quality in factories insulated from the rigours of the weather outside. The concept parallels the mass production of modern cars.

Assembly, on site, might be carried out with fixtures and computer-controlled robots making welded or mechanical joints to complete the assembly and pass-off test verification. Space capsules need very effective heat insulators that must be very light in weight. Structures have been devised using very thin materials in a way that achieves the desired result. Homes could be built, using similar designs for modular panel construction, but made of cheaper materials, as weight is less critical. Steel or plastic would be used in place of titanium used in space capsules. Visible panels could have ceramic or other coatings to suit both performance and aesthetics. It is crucial for new factory-made homes to be functionally excellent, affordable, and a source of pride in ownership. Citizens must want to live in such buildings or we will never achieve the cost and environmental benefits. There are some very appealing homes made using these faster building processes. Such buildings can embody local power generation and energy storage as an integral part of the building. The standardized modular approach built in high numbers with much automation is crucial to getting the quality right at the right price.

No car would be affordable if each one was made by hand, so why do we build homes by hand in all weathers? Avoiding a climate disaster requires better homes that are energy neutral and price supportable out of earned income.

Designing better communities #2: Resort-style living

Resort-style living might raise the quality of living and reduce costs. Once experience with this style of living has been sampled, it is likely to become highly desirable, especially for young workers with families. It should also meet the needs and the wishes of retired people.

Each home would have only what was needed for everyday living with no spare rooms or garages. The recent lockdown caused by the need to stay at home has raised the importance of each home having some outside space, with greenery where people from home can relax, and where children can play in the open air. Everything needed for occasional use would be shared and paid for on a per-use basis plus a standing access fee. Personal transportation would be done using a "grab taxi service". This allows residents to choose a vehicle that is well suited to the specific task for just the duration of that task. The vehicles may be driverless, or with a driver and a mate, as required by exceptionally onerous tasks, such as heavy deliveries. There would be shared meeting points to encourage people to meet and get to know and trust each other, and sports facilities.

There would be gardens tended by volunteers, or employed staff, with access to expertise. Work exchange or whatever the community favours as a fair way of achieving mutual support.

Because the shared facilities will be shared by many, perhaps including non-residents, who will pay their way, and are compatible with the resident community, the range of facilities can be higher than in a conventional home. They can also be of high quality and well maintained while costs are kept at supportable levels from earnings within the community. Not everyone will want the full package.

The concept is compatible with having a local delivery and return centre within the resort. There must be good connections with intercity services such as trains, buses, and planes. Clearly, there must be easy access to schools, hospitals, and all the other supporting services.

Most such communities will need their own plant growth factory. Extra trees should be planted using a wide range of trees and shrubs that support the wildlife needed. Those living in these communities should be encouraged to work locally, including those who work in the global workspace. Personal and face-to-face contact is vital, but it does not need to be all day, and every working day, when technology allows us to work with almost anyone living on our planet.

Many such communities might have their own direct air capture of CO_2 business that also provides for local liquid fuels for transport and hydrogen for domestic heating for powering buses and trucks, and local tram or rail services.

Funding the cost of affordable homes

Consider the case of office workers commuting daily from the only place they can afford to live to their office in a big city. For many, this takes up to four hours per day, every working day. By using technology most such workers have on their mobile phones, tablets, or laptop computers, they could reduce the time commuting by 90% and save thousands in travel costs.

Reduced commuting requires a re-think in the way fares are structured, and how travel loads are better balanced to get the right match of capacity, timing, and passenger numbers.

Using the UK as a worked example – pre-pandemic, most people travelled from home to intercity station by car, often incurring considerable annual parking costs (in the UK these average out at about £1000/year). Many use a big family car, while there is a second car back home for use by other family members. Car running costs are likely to exceed £2500.

Because the time away from home is so high, extra costs are incurred hiring support at home when there are young children or frail grandparents to care for. Taken together, a typical daily commuter could spend up to £10,000 per year.

The average cost in the UK of a small family house in 2019 is £251,000.[4] If we adopt resort-style living, a basic home with only the essentials, albeit to a good standard, and we use high volume factory-built modular construction, the new basic home will cost less than £100,000. We must add the cost of shared and maintained assets and facilities. This might add another £10,000 per year, making a total of £110,000. A 20% deposit drops from £50,000 to £22,000, a saving of £28,000. The amount requiring a mortgage drops from £200,000 to £88,000, a saving of £112,000.

By giving up the daily commute, many will have an extra £10,000 per year to help cover the reduced cost of buying/renting their own home. Moreover, they will have up to half a working week extra time per week to spend with family or take care of extra jobs. For example, many will want to help young people prepare for a good and rewarding adult life.

There is an acute need for this. Failure to provide such support has high social, health, and financial costs. As noted earlier, there are real issues that need to be addressed for working from home. Early research indicates people are working an extra week every month. There are issues relating to health and happiness, as people adjust to new ways of living and working. Many people will feel threatened by being less visible to colleagues and management. Effectiveness in working will shift from time management to the management of good quality output.

The new homes and resort facilities help reduce the carbon footprint substantially. The new homes will have minimum heat losses and emissions. They will be built from low carbon materials and construction methods. They will be well-insulated and optimized to minimize heating and air-conditioning costs and emissions. Many homes will have spaces well suited to productive work and have the right technology for team working with dispersed co-workers. The communities will harvest their energy needs from renewable sources via energy storage systems that match demand with some surplus/intake to ensure balance.

Notes

1 Legal and General Modular Homes (2021), *Creating Thriving Communities that Help to Solve Britain's Housing Crisis*. Available at https://www.legaland-general.com/modular/ (Accessed 8 March 2021).

2 World Economic Forum (2019), *This Start-Up is 3D-Printing an Entire Neighbourhood in Mexico*. Available at https://www.weforum.org/agenda/2019/12/3d-printed-homes-neighborhood-tabasco-mexico/ (Accessed 8 March 2021).

3 New Story (2019), *Press Kit*. Available at https://newstorycharity.org/press-kit/ (Accessed 11 March 2021).

4 GOV.UK (2020), *UK House Price Index England: November 2019*. Available at https://www.gov.uk/government/statistics/uk-house-price-index-england-november-2019/uk-house-price-index-england-november-2019 (Accessed 8 March 2021).

Part 6

UNHEEDED WARNINGS AND CRITICAL ACTIONS

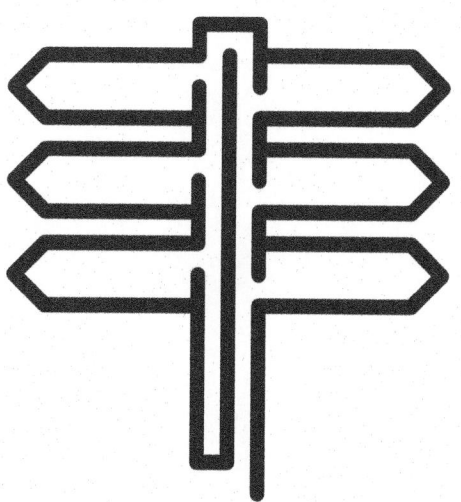

Critical actions for governments

- Empower innovation
- Provide financial incentives and seed funding to kick-start innovation
- Active engagement with citizens to avoid a climate disaster

DOI: 10.4324/9781003193470-19

14

UNHEEDED WARNINGS

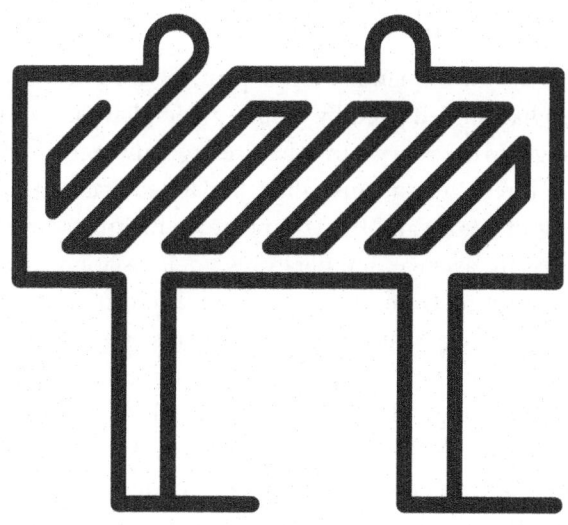

DOI: 10.4324/9781003193470-20

As an experienced aeroengine designer, I have faced challenges arising from problems we believed did not exist because we thought that we had used the best possible design methods, databases, and confirmatory tests. In these situations, you know that there may be something beyond your knowledge that will expose a weakness. Well-directed testing may reveal the true cause, but in some cases, it does not. In these circumstances, the designer must re-analyse the design and every possible operational procedure and make a list of every conceivable cause of the problem. The designer must find a solution for all these suspect features and operating procedures, and implement them all, provided everyone passes the following tests. Does this change make things better and not introduce a risk of making things worse? Often, the required degree of change is quite small.

The key point is that for many changes, we have an incomplete understanding of all the natural science involved. Moreover, the change may be used in ways that the designer could not have foreseen. If we wait until everything is known and understood, climate disaster will have arrived! Clearly, this is not acceptable. We must make everything safe despite not having a complete understanding. Engineers have done this for centuries. As population increases, combined with increasing power to spend on whatever we choose, we need ever more powerful tools to reduce the chance of making things worse. Nature has made it possible for the collective population of Earth to find, refine, and use powerful tools that can help secure sustainable human life, free from a climate disaster.

Has the tipping point for climate change arrived?

Before we look at a recommended action plan, let us recall all the warnings about climate change, delivered by nature, which have not resulted in the powerful changes to the way we live that are urgently needed. These include uncontrollable forest fires in Australia, the United States, and elsewhere, severe flooding in Venice and the UK, polar ice melting at ever-increasing rates, increases in average temperatures, deforestation that massively reduces CO_2 absorption from the atmosphere, and blanching of coral, plus the removal of peat bogs that store huge amounts of greenhouse gases.

Now we have COVID-19, which coupled with our global travel and the movement of food and goods, enabled it to quickly spread across the globe. We can clearly see and feel the impact of COVID-19, not just because of the horrific consequences but also in the brilliant actions being taken on a huge scale around the world. The challenge is to help people understand that climate change has the potential to kill many more people than pandemics like COVID-19, albeit in a less visible way initially. Despite the significantly larger threat, climate change has, to date, only provoked much weaker corrective actions.

With both COVID-19 and climate change, we are short of a clear understanding of important details, but there is much that we can do, and we can measure the effectiveness of our actions, and adjust the actions, in light of the new experience.

The global lockdown of industry and travel has given rise to a much cleaner atmosphere, making it clear that human activities are responsible for the degradation of the world's climate. We now have many examples of climate damage and clear evidence that when humans stop travelling and many other practices that damage the climate, the atmosphere responds positively.

Mother Nature is making it abundantly clear that many of the actions to limit the loss of life from coronavirus are also needed to stop climate change eating away at life on Earth for humans and other creatures. While we can streamline the severe actions taken to address coronavirus, we must reset the way humans live and work to achieve sustainable human life on Earth.

From the research done in collecting reliable material from which to write this book, it is clear that we urgently require a big reduction in producing greenhouses gases from all human activities. Actions taken to try to reduce the death toll from COVID-19 reveal some changes that are welcome as they improve the way we live, work, and play.

There are funds available with global investors making negative returns while awaiting a better investment opportunity. These funds will have been reduced by the lockdowns that have severely restricted economic output across the world, but we must use the resources we have to tackle climate change as we exit from the virus pandemic, and transition to new and better ways of living, working, and playing. The recovery needs a huge uplift in economic activity. Resourcefulness in lockdowns has supported the belief that we can do so much better. Moreover, despite lockdowns, there has been a huge increase in mutual support and respect for others. Climate change mitigation provides an unprecedented demand for new jobs in viable businesses that can benefit the UK, and many other countries, in terms of inward investment, national self-sufficiency, and improved living standards for every citizen. Special efforts must be made to ensure that everyone has access to technology, and the training to use it well. We must think in terms of providing every child with computers and skills. This is far more important than new infrastructure that panders to a Victorian style of living, which is no longer appropriate to success in a demanding global marketplace. Every country must strive for all its citizens to live well within the nationally earned income. Technology, used well, makes that achievable. Returning to the way we lived before the COVID-19 pandemic guarantees failure in reaching that goal. Moreover, the spiral down in living standards will hurt, with the disadvantaged feeling the hurt most severely. This may lead to unrest, adding greatly to national decline.

Our response to COVID-19 has shown us that we can co-operate and be courteous to others when lives are at risk. We're going to need this again, and on a larger scale and for even longer in order to alleviate the climate change threat to a good lifestyle for everyone. We all must make changes, but they can and must be changes for a better way of living.

People, especially younger generations, recognize the urgency and are angry that they see little happening. National governments are being pressed to set ambitious targets, but, despite their high level of competence, few appear to know enough to be able to set ambitious targets, which are based upon actions already demonstrated that they know will make a difference. There are understandable short-term economic gains by continuing to burn huge amounts of fossil fuel, take luxury cruises, and more, but doing so makes climate change worse and recovery from that damage even more challenging, if not impossible.

There is good evidence that we know many of the powerful things we can do across the globe to retain a healthy planet fit for sustainable human life of rising quality. But unless we move fast in attaining helpful actions on a global mass-market level, the runaway decline will accelerate.

15

CRITICAL ACTIONS

DOI: 10.4324/9781003193470-21

There are a few key principles that I think we can use as tools to help us make better decisions. We have discussed all of them at various points in the book, but I thought it might be helpful to set them out here:

- We should seek to harness technology effectively, both to reduce existing climate damage as well as preventing further damage.
- We must make, service, and grow locally, globally, creating jobs.
- We must make all energy sources green and accelerate the transition to green solutions. Specifically, this means the adoption of fuel cells and the acceleration of the move to a hydrogen-based economy.
- Existing assets must be converted to carbon-neutral fuels or removed from service.
- We must greatly reduce our dependency on foods produced by animals and expedite the creation and consumption of plant or cell-based alternatives.
- We must share our resources, talent, and actions and these must be supported by governments on a local, national, and global scale. Much of what we need to do requires funding and so we must re-balance national taxation to balance income with expenditure.

Mother Nature has done a brilliant job so far in taking care of human needs, but as world population and personal spending power have increased massively and nature's warnings gather pace and intensity, we still need to do more. We must not let human rights allow people the freedom to kill family and friends by ignoring sensible rules to prevent a climate disaster. Abiding by the rules gives everyone freedom. Citizens in strictly controlled Singapore have more freedom to live a full life in a safe clean environment than citizens in nations that allow unruly behaviour. Successful democracies depend upon rules of law, with full compliance.

Collectively, leading engineers have the skills necessary to pave powerful ways forward using the best-proven science and understanding of all the pertinent issues. We need our engineers to design new products that are cleaner to make and use. We also need to provide guidance on better ways to live and work that citizens can understand. Hopefully, this will enable everyone to adjust to life with lower emissions. While it is for the people to choose to live better, we need our engineers to highlight safe and proven products and services that will consistently help in alleviating climate change. Through our education and training, experts increasingly understand how nature works, and how best to eliminate the problems restricting output. Engineers are not required to replace nature. Nature just needs support now the task has risen so sharply through a sharp increase in population and human spending power.

Nature has enabled humans to collectively find the tools. Now is the time for humans to use these tools. Large-scale action across the globe is crucial to the mitigation of climate change. Nothing less will do. The developed nations must take leading roles while recognizing that some of the most powerful developments can come from anywhere. Innovation, engineering, manufacturing, management, vision, and financial skills must be used to grow national economies, self-sufficiency, and jobs, leading to an improved lifestyle provided out of earned income. Too high a dependency on borrowing results in both national and personal insecurity.

What are the actions?

Actions must be taken by people, noting that the 2019 world average green-house gas release per person per year is about 5 tonnes. UK citizens contribute about 5.5 tonnes, while citizens of the United States emit 16 tonnes.[1] These low figures reflect the export of jobs, and with them, a substantial portion of the tax base, thereby increasing the national need for borrowing. As we've seen, the biggest changes must come from the industrialized nations. Accordingly, those citizens must make the biggest change. The change must be made by everyone. Anything less will not avoid the impending climate disaster, with a heavy loss of life. For this to happen, everyone needs to understand what is at stake and the changes must make life better. Moreover, the changes must be affordable, and the associated businesses must generate returns in both the short and long term on a mass-market scale. Getting to bankable businesses will often need some government aid to get through the initial high-cost launch phase that precedes the attainment of mass-market volumes that can be delivered sustainably.

Clean energy

Maximize renewable power

Every nation needs to move to much-increased renewable power sources in order to clean up power from energy distribution networks while adding generating capacity to compensate for the withdrawal of fossil fuels. Solar and wind turbines, both offshore and onshore, will be the main providers, but this is only possible when backed by affordable energy storage. Much of that stored energy must be transportable from generator to consumer in a viable business structure. We have seen that we can fly our planes and run our cars and trucks on renewable energy, but first it must be transformed into the appropriate high energy density form. Importantly, we can choose a green high energy density fuel, such as hydrogen.

Move to a global green hydrogen-based economy

Earth is endowed with an abundance of hydrogen, but most of it is locked away in other materials. Hydrogen has the highest energy per unit weight of all the practical fuels used for power and heat. It burns cleanly, releasing only water as it delivers its heat. As the world's best escape artist, hydrogen places tough demands in providing affordable clean power to get into the right form for viable commercial use. Collectively, we understand how to achieve this.

Much of industry uses huge amounts of heat and power taken from the electricity and gas distribution grids. For many, the changeover is practical, provided the price per unit energy delivers results in products and services at realistic prices. Domestic heat and power from hydrogen are also realistic once there is a dependable and affordable supply network. A hydrogen gas grid for Europe, including the UK, is needed as soon as possible.

Note that for many power applications, fuel cells are more efficient than engines based upon the combustion of fuels. For many forms of transportation, an electric driveline may give rise to other valuable benefits.

Make nuclear power smaller and more affordable

Many industrialized nations will find that renewables plus energy storage that is transportable will require energy sourced from distant renewable energy farms. This raises the risk of power outages that may be caused by natural hazards or by international hostilities. The world is rarely free of tensions that can erupt with short notice. Nuclear power generation as a significant power generator in the mix has real merit in terms of energy security. However, large-scale nuclear generates power at too high a cost. Small-scale modular nuclear power generators built in large numbers might achieve a cost rate that enables industry to be cost competitive in the global marketplace.

For this to succeed many "launch customers" are needed to commit to taking and operating nuclear power stations, each with units having a peak output of about 300 MW. The modular stations would be built in factories, to a standardized design, in sufficient numbers to get the unit cost down. Moreover, there needs to be a standardized procedure for installation on site and the operation of the plant. It needs international co-operation. It needs a fast production timescale, such that the investment starts to generate revenue to repay the heavy up-front cost in the shortest possible timescale. Safety is assured through a very thorough design process, using the best technology and supported by tests. Safety following the onset of any problem can be assured by safe procedures.

Industries to move to renewables plus energy storage

Avoiding a climate disaster require a large contribution from all industries. Some will find renewable energy plus energy storage a viable choice. Others will find hydrogen for heating in heat-intensive industries the best choice.

Maximize every resource (natural and manufactured) capable of absorbing CO_2

CO_2 persists in the atmosphere for thousands of years. Trees and other plants require CO_2 to grow. Peat bogs are excellent absorbers of CO_2. CO_2 can be used to make building materials. We need to maximize our greenery and make sure our peat bogs remain wet and in full working order. Around the world, especially in Singapore, countries are using their high-rise buildings and roadside infrastructure covered with well-maintained trees, shrubs, and other greenery to absorb as much CO_2 as possible. The leaves of this greenery collect many of the fine particulates in a way that enables most of them to be disposed of safely. The greenery on buildings also provides shade within the buildings and reduces convective heat input to the atmosphere from buildings.

As well as increasing the resources for the natural absorption of CO_2, we must also accelerate and scale the innovations that enable the direct air capture of CO_2 from the atmosphere.

There are processes by which high power density liquid fuels can be made, starting with CO_2 captured directly from the air using renewable plus stored energy from renewables. The technology is known but there is a need for many plants sited strategically around industrialized countries, each with an output of one million tonnes per year.

Direct air capture of CO_2 technology is understood, and the first industrial unit capable of removing one million tonnes per year is being built now. We must use the CO_2 for a good purpose to make the scheme deliver tangible benefits. Progress is positive but there is considerable scope for increasing benefits while reducing costs.

Transport actions

Travel and transport less

Let the data and funds do the routine travel, thereby reducing travel and transport to close to the minimum necessary.

Utilize existing transportation with a long service life ahead

All industrialized countries have huge inventories of assets with a long potential economic life ahead. These must be adapted to run on affordable carbon-neutral fuels or hydrogen as fast as possible. Collectively, these assets are responsible for about 25% of the world annual output of greenhouse gas.

It should be possible to reduce this source further and faster than by replacing them all with electric power.

Electric cars need to be the norm

Electric cars are good for the environment, but they are currently too expensive to be serious contenders in helping prevent a climate disaster. The world needs more affordable electric cars. We also need electric cars that are better suited to peoples' needs if we are to achieve a global market uptake that will help avoid a climate disaster. We need smaller and lighter electric vehicles tailored to the urban usage-only market. We also need lighter single/twin-seat vehicles that fit the worker/commuter market often served by lightweight motorcycles. There is scope for lightweight and low drag forms with enough performance to run with the traffic while having full weather protection. The global market might approach one billion, built over a ten-year period. Such vehicles would need only a small traction battery and a small fuel cell. At the low end of this class of vehicles might be slow, pedal-assist models that could be ridden/driven by people aged 14 and upwards.

Existing long-range commercial aviation and other transportation must have a sustainable liquid fuel

Long-range commercial aviation is vital for world peace and trade. Avoiding a climate disaster requires commercial aviation to reduce its contribution. The only known way of doing this for the nearly new existing aircraft is to replace the fossil-fuelled high energy jet fuel with one of comparable energy but with a much higher carbon-neutral content. We must address the issue of greenhouse gas reduction for long-haul flights. Progress has been demonstrated, but we need to quicken the pace. Versions of carbon-neutral liquid fuels are potential fuels for cars vans and trucks, as the world transitions to electric- and hydrogen-fuelled vehicles. Brazil and France are pioneers in this movement. E85 in France needs to become part of the global vehicle fuelling infrastructure. In parallel, and with great urgency and thoroughness, commercial aviation must use hydrogen as fuel.

Food and agriculture

Eat less meat

Animals grown to meet the human desire for meat and other desirable food products are responsible for huge amounts of methane released into Earth's atmosphere. Methane is far more damaging than CO_2 as a greenhouse gas, even though it dissipates more quickly, it is the short-term hit that is of real concern.

Citizens of industrialized nations eat and drink more than is healthy, so people must trim their diet for health reasons, ideally by consent. If people are unwilling, then rationing may be necessary.

Informed diners in mass-market restaurants have shown a willingness to moderate their demand for meat in favour of an improved dining experience. Moreover, these diners are willing to try and enjoy meat and fish equivalents provided the taste, texture, consistency, and price are right.

Accelerate the mass production and adoption of meat-like foods

Work underway in creating synthetic meat needs to accelerate. Early results are impressive, but the products remain priced in the "luxury item" category rather than the "family staple" category. This will come in time, especially as innovations bring with them increasing competition and market penetration.

Consumers need to be encouraged to "take the first step", to reduce their meat consumption either by eating less meat or replacing some of their intake with synthetic equivalents.

Farmers using technology to improve yield, quality, and price competitiveness

Farmers must continue to embrace new technology, such as vertical farms, advanced environmental control, and speed breeding, which adopts new growing infrastructure such that we can grow more, with fewer resources, in places that are more accessible by those that will consume the produce.

Other ancillary technologies must also be explored and adopted to provide further benefits such as remote monitoring using drones to detect and deliver solutions such as animal distress, selective weed control, and efficient use of fertilizer.

Factory-grown food located close to communities

It takes much more land to feed a person than to house that person. This leads to high mileage in getting food from farms to supermarkets. We need communities with their own food factories and vertical farms that can provide year-round food crops of consistently high quality to match the predicted demand profile, at the right unit price for each product. These vertical farms avoid the use of pesticides, reduce water consumption by 90%, and greatly reduce wastage and food miles.

Homes and housing

Accelerate the move towards better, affordable, carbon-neutral housing

There are too many homes in industrialized countries that are poorly insulated, contributing a quarter of the greenhouse gases released by those nations. We must improve all that can be improved. Domestic power, heat, and air conditioning must use energy from green sources.

In many industrialized countries, there is a shortage of affordable housing, leading to excessively high prices for homes that are not well suited for living well in the years ahead. We need to greatly increase the build rate by using build techniques based upon new materials using new technology to provide for mutually supportive communities in green spaces. We must supplement traditionally built homes with modular homes made from cross-laminated timber, homes 3D printed from a concrete-like material, and stackable modular homes built in steel.

Government actions

Empowering innovation

Everyone needs access to technology and the skills to use it for innovation and access to global expertise.

We need common, shared resources that are easy to use to enable people to find the best practice to develop solutions that will help avoid a climate disaster.

Provide financial incentives and seed funding to kick-start innovation

Every country will need to revise how it pays for sustainable life out of earned income. Avoiding a climate disaster requires many changes, some of which reduce vital government revenue from taxes from previous sources. Consider road pricing in place of vehicle tax? Tax incentives are given in some countries to ease the cost of electric cars that cost more than fossil-fuelled cars because of the high cost of traction batteries. We may need tax incentives to encourage the purchase of more heat-efficient homes and factories. We may need a financial reward scheme for innovations that reach early production volumes. State-provided subsidies are required to get proven concepts properly financed as they pass from proof of concept to full-scale production for global markets. Multinational collaboration is often needed to reach the scale of uptake needed to significantly reduce the chance of a climate disaster. State intervention and state funding contributions may be needed in many cases.

Climate change demands the fastest possible attainment of global uptake of all projects that help avoid a climate disaster. Too many commercial investment funds are making negative returns, so we need to deploy these funds to finance the potentially viable projects that help avoid a climate disaster.

Active engagement with citizens to avoid a climate disaster

As the global population grows towards a likely peak of about 11 billion, sustainable human life requires the absence of human activities that accelerate the onset of a climate disaster. Just as we have seen with the "battle of information" surrounding mass vaccinations, appropriate engagement is needed to ensure all citizens are aware of their role in the solution of this global problem in order that human life of quality is sustainable. This needs to be widely understood, as good human behaviour by everyone is essential.

Final thoughts

There must be a plan for life on Earth to continue smoothly while the threat of a climate disaster is removed. All parts of the solution must be engineered to the highest professional engineering standards.

The plan must have executive managers to control every action essential for the avoidance of a climate disaster that will take far more lives than the COVID-19 pandemic, and from which there is no place for humans to hide. Humans share one atmosphere. Governments across the globe must work together with all their resources and expertise to retain sustainable life for all. The plan must be regularly monitored to ensure every part of the plan will be delivered on time and in full. Any slippage must be recovered in the plan. The time horizon must initially be 2030, to be revised as progress is achieved. Proven early progress gives us more time. Failure to progress shortens the time and reduces the resources for action.

There is more than enough evidence to know that a climate disaster is avoidable, provided all the right actions are fully implemented across every community. More than this, all actions improve the quality of life in a way that is supportable out of earned income.

The views of climate change deniers are important. They help to clarify the need for action.

Nations and human rights activists must understand that for eight billion humans sharing a single atmosphere, there must be codes of behaviour to prevent the actions of a few resulting in many deaths. There must be globally visible encouragement for those who can lead the way and inspire others to change their behaviour.

Everyone with knowledge must engage with others, wherever they are, to get the volume of effective action up to the scale needed to avoid a climate disaster. This process of levelling up must be near the top of the to-do list for every government. Recycling of pre-used computers must be part of this scheme. The provision of computers, and the training to use them, will encourage global conversations that will illuminate the urgent need for deliverable action on climate change. Moreover, it will unlock many more innovations.

We need to monitor the emergence of relevant innovations, and their uptake globally, making progress visible to everyone. The record needs to reveal the potential impact and the actual impact, showing how this is achieved.

We must use national budgets to discourage bad practice and encourage the introduction of innovative businesses aimed at reducing emissions at the necessary fast rate. National tax schemes repeatedly fail to have a long-term vision. Nations cannot repay their debt by taxing those without a job. Climate change requires new jobs, not old jobs extended beyond their point of business failure. Avoiding a climate disaster causes changes to where people live, work, and play. The support services must adjust or die. Government help needs to ease the processes of adjustment.

We must respond to the changing balance of international trading. The buying power of citizens in Southeast Asia, China, and India is rising faster than elsewhere. Moreover, they have better opportunities to produce climate-friendly goods and services. Many Eastern countries have aggressive programmes for smart cities, requiring a wide range of technological equipment not yet produced in volume. Will Western countries buy or sell what is needed? Will Western countries have the buying power?

The bottom line

Without a deliverable plan across the world, at the right scale and fast enough pace, climate change could be disastrous. It has started with floods, violent storms, wildfires, and other disasters nibbling away at human life and livelihoods. Year on year we can expect increasing hurt. But it need not be that way. Collectively, we know what actions to take, and we know it can be affordable out of earned income while raising the quality of every human life. There are no reasons for inaction, as the required changes make things better. People must reset the way they live, work, and play, then enjoy a better life.

Throughout my life, I have been mentored and inspired by many outstanding design engineers, business leaders, and academics. These men and women have made a difference in improving life for many through their thinking, powers of persuasion, and actions. Many have gained great honour for their contribution, but few got everything right, so those that follow have a huge potential to improve things further.

How did I get access to these wonderful people? My dad left school at 13 and worked hard to become a craftsman making wooden patterns from which new cars, motorcycles, and machine tools were made. My mother left school at 14 and took a job, and like my dad studied at night school to enable her to get better jobs. I was born with a hernia that caused great pain until I was five and old enough to withstand the operation. I started school one year late.

My inherited work ethic and a supportive mum and dad enabled me to catch up. Without realizing it, catching up from behind means that at the winning post you are going faster than the leaders. This gets noticed and opportunities follow, adding yet more to pace, experience, and width of opportunity.

Good schools, good universities, and membership of clubs and professional bodies offer opportunities to see and hear of innovative work. I have enjoyed my life, made possible by choosing to become a design engineer and marrying my wife with her caring and strong work ethic.

As I set out right at the beginning, I remain optimistic and enthused by everything that is possible on the road ahead. There has never been a better time for young people to become design engineers with the power to not just help avoid a climate disaster but to also raise the quality of life for every single person that shares our beloved planet.

Note

1 Global Carbon Project (2019), *Global Carbon Atlas.* Available at http:// globalcarbonatlas.org/ens/content/welcome-carbon-atlas   (Accessed   27 February 2021).

ACKNOWLEDGEMENTS

I thank the truly outstanding people for creating the possibility for me to write this book, and for massive support while harvesting information from sources spread out across the world, and to bring it all together in a way that helps everyone to understand the need for humans to enjoy better lives in ways that help avoid a climate disaster.

In December of 2017, I was on holiday with my eldest son, Steve, and his family living in Singapore. As I relaxed, it became clear that some parts of the world had recognized the threat of a climate disaster and had already implemented many small steps to remove greenhouse gas emissions. Indeed, some of the steps, like preparations for zero-emission aviation, were starting to appear. It was also clear that countries in Southeast Asia saw the avoidance of a climate disaster as the source of many well-paid jobs, much needed by large and fast increasing populations. This prompted me, as an innovative design engineer, to realize that there is only one atmosphere in which every human lives. Earth's atmosphere is complex, with many interactions. Designing an aeroengine presents the same sort of complexity, albeit on a smaller scale. This led to the thought that the view of an experienced design engineer might illuminate the complex network of pathways needed to avoid a climate disaster. As with any complex system with a potential for life-threatening surprises, these must not be allowed to take lives. Accordingly, the disciplines of a leading aeroengine designer fit well with seeking out the changes that help avoid a climate disaster. Our middle son, Richard, our tame mathematician and barrister, has a lovely knack of reducing every problem to a few equations. Resolving these equations quickly exposes who is responsible for each action. Clarity with accountability really does help focus the necessary action.

I started to plan and write, but within a few months, I learnt that I had five component failures in my heart, and the collective view of the experts was that surgery risks were too high and medication was decided upon. It worked like a dream for six months. Then my heart started to fail rapidly. At this point our second son, Richard, intervened, leading to our finding Dr Ramesh da Silva, consultant in interventional cardiology, who referred me to a truly outstanding heart surgeon, Mr Narain Moorjani, at The Royal Papworth Hospital. Within minutes of the first meeting, it was clear that his approach to life saving in the presence of a life-threatening situation was the one I used as an aeroengine designer, giving me confidence in a potential solution. He wrote to me with a plan, which was clear, but missing the rationale behind his plan, I challenged it, and he made it clearer. There were five days in intensive care while heavily instrumented to provide all the data needed for a long sequence of repairs to be carried out by the required number of surgeons, lasting most of a full working day. Mr Moorjani is also an author. He knows that giving birth to a new book is demanding. As I emerged from the anaesthetic, I knew I had a heart that was in good order. I express my sincere thanks to Mr Narain Moorjani, his team, and the support of the Royal Papworth Hospital, without whom this book would not have been written. The team at The Royal Papworth depend upon innovation, delivered with thoroughness in making sure that risks associated with innovation are contained with the highest possible standards of safety.

I thank my son Steve who is senior vice president for manufacturing in a large UK company, heavily involved in providing antiseptic materials across the world. He has responsibility for many factories across the world. Conversations with Steve give me a window into developments in many countries with huge cultural differences. During the coronavirus pandemic, he has built new factories in distant lands and got them running with competent staff, and some very sophisticated machinery, while not being allowed to travel. Digital technology has not only made things possible during lockdown but it has illuminated better ways of working that need to be carried forward when the COVID-19 pandemic is controlled. Thanks Steve and thanks for the support from your family.

Changes to make things better take the changemakers into danger. Thanks must go to my parents who taught me to use dangerous tools safely. This is so relevant to the changes we all must make to avoid a climate disaster and increase the standard of living for everyone.

My school, Bablake in Coventry, provided pupils generous opportunities outside the walls of the school to see the world of work. Moreover, the headmaster brought in experts breaking new ground to illustrate the excitement when pioneering changes lead to improved ways of living, working, and playing. I joined the school, aged ten, immediately after the end of World War 2 in Europe and was able to use my skills, working safely with dangerous woodworking tools, to help return the school classrooms from dormitories back to classrooms, uplifted with prints of famous paintings all framed by me. All this is relevant as we must ensure safety, as we all transition to a better life, that is different from what most people are used to. Thank you to Bablake School.

My university, Imperial College in London, where I studied aeronautical engineering, is situated in one of the world's great centres where learned papers are discussed, covering every branch of science.

I thank my son David for guiding me in using some of the most powerful digital tools to help me evolve my thinking and to help me transfer the results into the words you see on the page in front of you. It is a real joy to work with Dave on a project that we both see to be important in helping everyone to understand that there are ways to avoid a climate disaster and that the changes will make life better. The political challenge is that of making sure it is better for everyone.

Most of all, I thank my wife Jean, to whom I have been married for over 63 years, for her love and constant support.

INDEX